MEMOIRS
of the
American Mathematical Society

ISSN 0065-9266 (print)
ISSN 1947-6221 (online)

Number 1497

On Singularity Properties of Word Maps and Applications to Probabilistic Waring Type Problems

Itay Glazer
Yotam I. Hendel

July 2024 • Volume 299 • Number 1497 (fourth of 7 numbers)

AMERICAN MATHEMATICAL SOCIETY

Library of Congress Cataloging-in-Publication Data

Cataloging-in-Publication Data has been applied for by the AMS.
See http://www.loc.gov/publish/cip/.
DOI: https://doi.org/10.1090/memo/1497

Memoirs of the American Mathematical Society

This journal is devoted entirely to research in pure and applied mathematics.

Subscription information. Beginning in 2024, Memoirs will be published monthly through 2026. Memoirs is also accessible from www.ams.org/journals. The 2024 subscription begins with volume 293 and consists of twelve mailings, each containing one or more numbers. Individual subscription prices for 2024 are as follows. For electronic only: US$1013. For paper delivery: US$1,163. Add US$22 for delivery within the United States; US$150 for surface delivery outside the United States. Upon request, subscribers to paper delivery of this journal are also entitled to receive electronic delivery. For information on institutional pricing, please visit `https://www.ams.org/publications/journals/subscriberinfo`. Subscription renewals are subject to late fees. See `www.ams.org/journal-faq` for more journal subscription information. Each number may be ordered separately; please specify number when ordering an individual number.

Back number information. For back issues see `www.ams.org/backvols`.

Subscriptions and orders should be addressed to the American Mathematical Society, P. O. Box 845904, Boston, MA 02284-5904 USA. *All orders must be accompanied by payment.* Other correspondence should be addressed to 201 Charles Street, Providence, RI 02904-2213 USA.

Memoirs of the American Mathematical Society (ISSN 0065-9266 (print); 1947-6221 (online)) is published bimonthly (each volume consisting usually of more than one number) by the American Mathematical Society at 201 Charles Street, Providence, RI 02904-2213 USA. Periodicals postage paid at Providence, RI. Postmaster: Send address changes to Memoirs, American Mathematical Society, 201 Charles Street, Providence, RI 02904-2213 USA.

This publication is indexed in *Mathematical Reviews*®, *Zentralblatt MATH*, *Science Citation Index*®, *Science Citation Index*™*-Expanded*, *ISI Alerting Services*℠, *SciSearch*®, *Research Alert*®, *CompuMath Citation Index*®, *Current Contents*®*/Physical, Chemical & Earth Sciences.*
This publication is archived in *Portico* and *CLOCKSS*.
Printed in the United States of America.

♾ The paper used in this book is acid-free and falls within the guidelines established to ensure permanence and durability.
Visit the AMS home page at `https://www.ams.org/`

10 9 8 7 6 5 4 3 2 1   29 28 27 26 25 24

Contents

Chapter 1. Introduction 1
Acknowledgments 2
1.1. Motivation 2
1.2. Main results 5
1.3. Further discussion of the main results 9
1.4. Methods and main ideas of the proofs 10
1.5. Conventions 14

Chapter 2. Preliminaries 17
2.1. Basic properties of jet schemes 17
2.2. Log canonical threshold 18
2.3. A method of degeneration 19

Chapter 3. Some properties of the algebraic convolution operation 23
3.1. The algebraic convolution operation: The case of general morphisms 23
3.2. Singularity properties of schemes and morphisms via their jets 25
3.3. Singularity properties of word maps: First observations 28

Chapter 4. Lie algebra word maps 31
4.1. Definitions and discussion of Lie algebra word maps 31
4.2. Formulation of the main theorems 32
4.3. Combinatorial picture in the case of the commutator word 33
4.4. Reduction to self-convolutions of homogeneous word maps of pure type 36
4.5. Combinatorial description of homogeneous word maps 38

Chapter 5. Lie algebra word maps: Proof of the main theorems 45
5.1. Discussion of the main methods 45
5.2. Lie algebras of low rank and small polyhypergraphs 50
5.3. Proof for $\mathfrak{sl}_n$ 51
5.4. Matrix word maps 57
5.5. Proof of the general case 59

Chapter 6. The commutator map revisited 67

Chapter 7. Local behavior of word maps and a lower bound on the log canonical threshold of their fibers 71
7.1. The degree of a word and of a word map 71
7.2. Algebro-geometric results on word maps 73

Chapter 8. Number theoretic interpretation of the flatness, ε-jet flatness and (FRS) properties 77
8.1. The Lang-Weil bounds and a number theoretic characterization of flatness 77
8.2. Number theoretic and analytic interpretation of the (FRS) property 79
8.3. The Denef-Pas language and motivic functions 80
8.4. Number theoretic characterization of ε-jet flatness 82

Chapter 9. Applications to the p-adic probabilistic Waring type problem 87
9.1. Algebraic random walks: General case 87
9.2. Algebraic random walks induced from word maps 91

Bibliography 95

Abstract

We study singularity properties of word maps on semisimple Lie algebras, semisimple algebraic groups and matrix algebras and obtain various applications to random walks induced by word measures on compact p-adic groups.

Given a word w in a free Lie algebra $\mathcal{L}_r$, it induces a word map $\varphi_w : \mathfrak{g}^r \to \mathfrak{g}$ for every semisimple Lie algebra $\mathfrak{g}$. Given two words $w_1 \in \mathcal{L}_{r_1}$ and $w_2 \in \mathcal{L}_{r_2}$, we define and study the convolution of the corresponding word maps $\varphi_{w_1} * \varphi_{w_2} := \varphi_{w_1} + \varphi_{w_2} : \mathfrak{g}^{r_1+r_2} \to \mathfrak{g}$.

By introducing new degeneration techniques, we show that for any word $w \in \mathcal{L}_r$ of degree d, and any simple Lie algebra $\mathfrak{g}$ with $\varphi_w(\mathfrak{g}^r) \neq 0$, one obtains a flat morphism with reduced fibers of rational singularities (abbreviated an (FRS) morphism) after taking $O(d^4)$ self-convolutions of φ_w. Similar results are obtained for matrix word maps. We deduce that a group word map of length ℓ becomes (FRS), locally around identity, after $O(\ell^4)$ self-convolutions, for every semisimple algebraic group $\underline{G}$. We furthermore provide uniform lower bounds on the log canonical threshold of the fibers of Lie algebra, matrix and group word maps. For the commutator word $w_0 = [X, Y]$, we show that $\varphi_{w_0}^{*4}$ is (FRS) for any semisimple Lie algebra, improving a result of Aizenbud-Avni, and obtaining applications in representation growth of compact p-adic and arithmetic groups.

The singularity properties we consider, such as the (FRS) property, are intimately connected to the point count of fibers over finite rings of the form $\mathbb{Z}/p^k\mathbb{Z}$. This allows us to relate them to properties of some natural families of random walks on finite and compact p-adic groups. We explore these connections, characterizing some of the singularity properties discussed in probabilistic terms, and provide applications to p-adic probabilistic Waring type problems.

Received by the editor August 23, 2021, and, in revised form, December 17, 2021.

Article electronically published on July 11, 2024.

DOI: https://doi.org/10.1090/memo/1497

2020 *Mathematics Subject Classification.* Primary 14B05, 20G25, 20F69, 17B45, 60B15; Secondary 11P05, 14B07, 03C98, 14G05.

Both authors were partially supported by ISF grant 249/17, BSF grant 2018201 and by a Minerva foundation grant.

CHAPTER 1

Introduction

In this paper we study word maps on semisimple Lie algebras, semisimple algebraic groups, matrix algebras, and compact p-adic groups.

Let w be a word, i.e. an element of a free group F_r, a free Lie algebra $\mathcal{L}_r$ or a free associative algebra $\mathcal{A}_r$, all taken over a set of r elements. The word w induces a word map φ_w when applied to an algebraic group, Lie algebra or matrix algebra, respectively. We focus on the singularity properties of such word maps, where an interesting phenomenon occurs; if one takes successive self-convolutions $w^{*t} := w * w * \cdots * w$ of w, then the singularities of $\varphi_{w^{*t}}$ improve as t grows. Here, convolution just means concatenation (resp. summation) of group (resp. algebra) words in distinct variables.

In [**AA16**], the commutator word $w = [X, Y]$ was investigated and it was shown that after 21 self-convolutions, the word map $\varphi_{w^{*21}}$ has mild singularities (it is flat, with fibers of rational singularities) when applied to any semisimple algebraic group or Lie algebra. In this paper we generalize this result to arbitrary word maps (Theorems A, B, C, D and E, Chapters 4, 5 and 7).

While the commutator map has a simple description in terms of polynomial equations, arbitrary words can be very complicated. In particular, Lie algebra word maps tend to have wild behavior when applied to simple Lie algebras of rank which is comparable to or smaller than the degree of the word. This makes it a-priori not clear how to utilize degeneration techniques in a meaningful way, as was done in [**AA16**]. We deal with these difficulties by introducing new degeneration techniques, as well as a systematic way to execute them (Chapter 5.1), and using our previous results [**GH19, GH21**] to deal with the situations where the word maps are wild. Using these methods, we were also able to give a simple alternative proof of the commutator result of [**AA16**], with improved bounds (Chapter 6 and Theorem F). New connections and reductions from group words to (Lie) algebra words are presented as well (Chapter 7).

In the setting of compact p-adic groups, we study the family of random walks induced by word measures, and obtain various uniform bounds on L^∞ and L^1 mixing times, as well as uniform upper bounds on the probability densities (Theorems I, J and L, Chapters 8-9). Similar questions have been recently investigated in [**LST19**] in the realm of finite simple groups. The main obstacle in the setting of compact p-adic groups, is that their representation theory is more complicated and far less understood than of finite simple groups, so representation theoretic techniques are less effective. Furthermore, here the singularities of the word maps control the size of the fibers, which requires the use of more advanced algebro-geometric tools.

There is a tight connection between the p-adic setting and the algebro-geometric setting described above. Thus, in order to deal with the p-adic setting, we provide new analytic and arithmetic characterizations of various singularity properties, which allow us to translate our algebro-geometric results to a probabilistic language. To do this, we incorporated recent advancements in the theory of motivic integration (e.g. **[CGH18]**). This allowed us to provide an algebro-geometric characterization of the property of uniform L^∞-mixing time (Theorem G), and further to relate between lower bounds on the log canonical threshold of fibers of word maps, to upper bounds on the density of word measures on compact p-adic groups (Theorem H).

Acknowledgments

We thank Uri Bader, Itai Benjamini, Joseph Bernstein, Raf Cluckers, Persi Diaconis, Ronen Eldan, Tsachik Gelander, Dmitry Gourevitch, Boris Kunyavskii, Gady Kozma, Erez Lapid, Michael Larsen, Martin Liebeck, Dan Mikulincer, Doron Puder, Aner Shalev, Shai Shechter, Stephan Snegirov and Pham Huu Tiep for useful conversations and for commenting on various preliminary versions of this project. We thank Nir Avni for many useful conversations and discussions, and for answering various questions. Finally, we wish to thank Rami Aizenbud for countless useful conversations, and for his constant guidance and support throughout this project. Part of this work was done during the "Group Representation Theory and Applications" program at MSRI.

1.1. Motivation

1.1.1. Waring type problems. Our motivation goes back to the eighteenth century, with Lagrange's famous four square theorem—every natural number can be represented as the sum of at most four square integers (Lagrange, 1770). Waring then considered the following generalization which was confirmed by Hilbert in 1909 **[Hil09]**:

QUESTION 1.1 (Waring's problem, 1770). *Can every natural number be represented as the sum of $t(k)$ many k-th powers, for some function $t(k)$ of k?*

We now formulate Waring's problem in a language suitable for our purposes.

DEFINITION 1.2. Let $\varphi : X \to G$ and $\psi : Y \to G$ be maps from sets X and Y to a (semi-)group G. Define the convolution $\varphi * \psi : X \times Y \to G$ of φ and ψ by

$$\varphi * \psi(x, y) = \varphi(x) \cdot \psi(y).$$

Furthermore, we denote by $\varphi^{*t} : X^t \to G$ the t-th convolution power (or self-convolution) of φ.

In this language, the Waring problem asks whether the power map $\varphi_k : \mathbb{N} \to \mathbb{N}$ defined by $\varphi_k(x) = x^k$ becomes surjective after $t(k)$ self-convolutions. Here is an interesting variant.

Given a word $w = x_{i_1}^{n_1} \cdot \dots \cdot x_{i_l}^{n_l} \in F_r$, with $x_{i_j} \in \{x_1, \dots, x_r\}$ and $n_j \in \mathbb{Z}$, and a group G, we associate the corresponding word map $\varphi_w : G^r \to G$ by $(g_1, \dots, g_r) \mapsto g_{i_1}^{n_1} \cdot \dots \cdot g_{i_l}^{n_l}$. Note that $\varphi_{w_1} * \varphi_{w_2} = \varphi_{w_1 * w_2}$, where $w_1 * w_2 \in F_{r_1+r_2}$ is the concatenation of $w_1 \in F_{r_1}$ and $w_2 \in F_{r_2}$. We can then ask if for any $w \in F_r$, there exists $t(w)$ such that $\varphi_w^{*t(w)} : G^{rt(w)} \to G$ is surjective for every G in a nice family $\mathcal{G}$

of groups. The minimal such $t(w)$ is called the $\mathcal{G}$-width of w. Similarly, one can take a word w in a free Lie algebra $\mathcal{L}_r$ on a finite set $\{X_1, \dots, X_r\}$ and consider, for any semisimple Lie algebra $\mathfrak{g}$, the corresponding Lie algebra word maps $\varphi_w : \mathfrak{g}^r \to \mathfrak{g}$.

This kind of questions are referred to as Waring type problems, and they have been extensively studied over the past few decades, in various settings. We mention some of them:

(1) $\mathcal{G}_{\mathbf{Alg}}$**, the family of simple algebraic groups.** It is a classical theorem of Borel [**Bor83**] (see also [**Lar04**]) that for any non-trivial word $w \in F_r$, and any semisimple algebraic group $\underline{G}$, the map $\varphi_w : \underline{G}^r \to \underline{G}$ is dominant. In particular φ_w^{*2} is surjective.

(2) $\mathcal{G}_{\mathbf{LieA}}$**, the family of simple Lie algebras.** In [**BGKP12**], an analogue of Borel's theorem was shown for semisimple Lie algebras $\mathfrak{g}$ under the additional assumption that $\varphi_w : \mathfrak{g}^r \to \mathfrak{g}$ is not identically zero on $\mathfrak{sl}_2$.

(3) $\mathcal{G}_{\mathbf{fin}}$**, the family of finite simple groups.** The problem of finding the width of various words over (large enough) finite simple groups has been studied extensively (e.g. [**MZ96, SW97, LS01, LS08, Sha09, LS09, LOST10**]). The state of the art is [**LST11**], where Larsen, Shalev and Tiep showed that for any $w \in F_r$, the word map $\varphi_w^{*2} : G^{2r} \to G$ is surjective for every large enough $G \in \mathcal{G}_{\text{fin}}$.

(4) $\mathcal{G}_{\mathbf{cp}}$**, the family of compact** p**-adic groups.** In [**JZ08**] it was shown that any word $w \in F_r$ has finite width with respect to any compact p-adic group. In [**AGKS13**], the authors showed that for any $w \in F_r$ and any simply connected semisimple $\mathbb{Q}$-group $\underline{G}$, $\varphi_w^{*3} : \underline{G}(\mathbb{Z}_p)^{3r} \to \underline{G}(\mathbb{Z}_p)$ is surjective for every large enough prime p.

(5) **Other families.** Given a ring R, one can consider the corresponding Waring problem for $\underline{G}(R)$ where $\underline{G} \in \mathcal{G}_{\text{Alg}}$. For example, the cases of $R = \mathbb{R}, \mathbb{Q}_p$ were treated in [**HLS15**], and the case $R = \mathbb{Z}$ and $\underline{G} = \mathrm{SL}_n$ was studied in [**AM19**], and it was shown that φ_w^{*87} is surjective for any $w \in F_r$, if n is large enough.

1.1.2. Probabilistic Waring type problems. Given that a word map φ_w becomes surjective after enough self-convolutions, the next step is to study its fibers $\varphi_w^{-1}(g)$, and to ask how many self-convolutions does it take so that the fibers of φ_w^{*t} will be uniform in some sense. Such questions are referred to as probabilistic Waring type problems, and have recently received a significant amount of attention (e.g. [**GS09, SS11, LS12, AA16, Bor17, LST19**]). In particular, one can take each of the Waring type problems described above and produce its probabilistic counterpart.

For $\mathcal{G}_{\text{fin}}$ and $\mathcal{G}_{\text{cp}}$, problems as above can be phrased in terms of questions on certain random walks. For simplicity of presentation we work here only with SL_n. Any word $w \in F_r$ induces a family $\{\varphi_{w,n,p,k}\}_{n,p,k}$ of word maps

$$\varphi_{w,n,p,k} : \mathrm{SL}_n^r(\mathbb{Z}/p^k\mathbb{Z}) \to \mathrm{SL}_n(\mathbb{Z}/p^k\mathbb{Z}).$$

One can then study the family of random walks induced by $\varphi_{w,n,p,k}$. Write $\pi_{n,p,k}$ for the uniform probability measure on $\mathrm{SL}_n(\mathbb{Z}/p^k\mathbb{Z})$, and set

$$\tau_{w,n,p,k} := (\varphi_{w,n,p,k})_*(\pi_{n,p,k}).$$

Then the collection $\{\tau_{w,n,p,k}\}_{n,p,k}$ induces a family of random walks on

$$\{\mathrm{SL}_n(\mathbb{Z}/p^k\mathbb{Z})\}_{n,p,k},$$

where the probability to reach $g \in \mathrm{SL}_n(\mathbb{Z}/p^k\mathbb{Z})$ at the t-th step of the random walk is

$$\tau_{w,n,p,k}^{*t}(g) = \frac{\left|(\varphi_{w,n,p,k}^{*t})^{-1}(g)\right|}{\left|\mathrm{SL}_n(\mathbb{Z}/p^k\mathbb{Z})\right|^{rt}}.$$

Since $\varphi_{w,n,p,k}$ is generating for p large enough (e.g. [**AGKS13**, Theorem 2.3]), $\tau_{w,n,p,k}^{*t}$ converges to the uniform probability $\pi_{n,p,k}$ in any of the L^a-norms for $a \in [1,\infty]$, as t goes to ∞. Roughly speaking, the minimal number of steps t_a it takes for $\tau_{w,n,p,k}^{*t_a}$ to be sufficiently close to $\pi_{n,p,k}$ in the L^a-norm is called the L^a-mixing time (see Definition 1.12). A priori, t_a might depend on p, k and n. It is natural to ask the following:

QUESTION 1.3. *Let $w \in F_r$ be a word.*

(1) *Is there $\varepsilon(w) > 0$, such that $\tau_{w,n,p,k}(g) < \left|\mathrm{SL}_n(\mathbb{Z}/p^k\mathbb{Z})\right|^{-\varepsilon(w)}$ uniformly in n, p and k, when p is large enough?*
(2) *Is there an upper bound on the L^a-mixing times of $\{\tau_{w,n,p,k}\}$ which is uniform in n, p and k, when p is large enough?*

1.1.3. Algebraic Waring type problems. Let $\varphi : X \to \underline{G}$ be a morphism from a smooth variety X to an algebraic group $\underline{G}$. Similarly to the above problems, we can study the convolution powers $\varphi^{*t} : X^t \to \underline{G}$ of φ using Definition 1.2. Similarly to the usual convolution operation in analysis, the algebraic convolution operation has a smoothing effect on morphisms. In [**GH19, GH21**], it was shown that if $\varphi : X \to \underline{G}$ is (strongly) dominant, then $\varphi^{*t} : X^t \to \underline{G}$ has increasingly better singularity properties as t tends to ∞ (see Section 3 for further discussion).

Now, given $w \in F_r$ (resp. $w \in \mathcal{L}_r$), we can study self-convolutions φ_w^{*t} of the corresponding word maps $\varphi_w : \underline{G}^r \to \underline{G}$ (resp. $\varphi_w : \mathfrak{g}^r \to \mathfrak{g}$) for $\underline{G} \in \mathcal{G}_{\mathrm{Alg}}$ and $\mathfrak{g} \in \mathcal{G}_{\mathrm{LieA}}$. The question of surjectivity of self-convolutions was fully answered by Borel's Theorem for $\underline{G} \in \mathcal{G}_{\mathrm{Alg}}$ (see [**Bor83**]), and partially answered in [**BGKP12**] for $\mathfrak{g} \in \mathcal{G}_{\mathrm{LieA}}$. Having the results of [**GH19, GH21**] in mind, and since there is a strong connection between the singularities of a variety and its point count over finite rings (see Theorem 1.21), we are led to the following questions, which are analogous to Question 1.3 (brought here in non-precise form):

QUESTION 1.4. *Let $w \in F_r$ (resp. $w \in \mathcal{L}_r$) be a word.*

(1) *How large are the fibers of φ_w, and what can one say on their singularities?*
(2) *Is there $t \in \mathbb{N}$ such that φ_w^{*t} has "nice" singularity properties, uniformly over $\underline{G} \in \mathcal{G}_{\mathrm{Alg}}$ or over $\mathfrak{g} \in \mathcal{G}_{\mathrm{LieA}}$?*

Question 1.4 is in some sense a generalization of Question 1.3. If we take

"nice"="flat, with geometrically irreducible fibers",

then the Lang-Weil bounds give a strong connection between Question 1.4 and the probabilistic Waring problem for finite simple groups of Lie type. Here, to measure how large are the fibers we use their dimension. This can already be seen in [**LST19**], where this connection is being used frequently. If we take

"nice" = "flat, with geometrically irreducible,
reduced fibers of rational singularities",

and consider the log canonical threshold $\mathrm{lct}(\underline{G}^r, \varphi_w^{-1}(g))$ (see e.g. [**Mus12**]), as a measurement of how bad are the singularities of the fibers, then Question 1.4

becomes strongly connected to the probabilistic Waring problem for compact p-adic groups.

In this paper we deal with Questions 1.3 and 1.4 in the cases of $\mathcal{G}_{\mathrm{Alg}}$, $\mathcal{G}_{\mathrm{LieA}}$ and $\mathcal{G}_{\mathrm{cp}}$. We answer Question 1.4 for semisimple Lie algebras, and provide partial answers for semisimple algebraic groups (Chapters 3, 4, 5 and 7). In Chapters 8 and 9 we study the connection between Questions 1.3 and 1.4, and utilize it in order to deal with Question 1.3 in Section 9.2. In addition, in Section 6 we use the methods developed in Chapters 4 and 5 to study the fibers of convolutions of the commutator map. As an application, we provide improved bounds on the representation growth of compact p-adic and arithmetic groups.

1.2. Main results

We divide the main results into several parts. We first discuss the relative Waring problem in the setting of semisimple Lie algebras and algebraic groups, through the study of singularity properties of word maps, followed by a discussion on the special case of the commutator map. We then provide algebraic interpretation to the various probabilistic properties of word measures in terms of these singularity properties. Combining these two parts, we provide applications to the p-adic probabilistic Waring problem.

1.2.1. Singularity properties of word maps.

Let K be a field of characteristic 0.

DEFINITION 1.5 (The (FRS) property [**AA16**, Definition II]).

(1) A K-variety X has rational singularities if it is normal and for any resolution of singularities $\pi : \widetilde{X} \to X$ we have $R^i\pi_*(\mathcal{O}_{\widetilde{X}}) = 0$ for $i \geq 1$.
(2) Let X and Y be smooth K-varieties. We say that a morphism $\varphi : X \to Y$ is (FRS) if it is flat and if every fiber of φ is reduced and has rational singularities.

DEFINITION 1.6. Let $\mathcal{A}_r$ (resp. $\mathcal{L}_r$) be the free associative K-algebra (resp. the free K-Lie algebra) on a finite set $\{X_1, \ldots, X_r\}$. We call $w \in \mathcal{A}_r$ (resp. $w \in \mathcal{L}_r$) an algebra word (resp. a Lie algebra word). Note that both $\mathcal{A}_r$ and $\mathcal{L}_r$ have a natural gradation. We define the degree of w, denoted $\deg(w)$, as the maximal grade $d \in \mathbb{N}$ in which the image of w is non-trivial.

In the following theorems we show that (Lie) algebra word maps obtain good singularity properties after sufficiently many convolutions, in a uniform way:

THEOREM A (Theorem 4.9). *Let $\{w_i \in \mathcal{L}_{r_i}\}_{i\in\mathbb{N}}$ be a collection of Lie algebra words of degree at most $d \in \mathbb{N}$. Then there exists $0 < C < 10^6$, such that for any simple K-Lie algebra $\mathfrak{g}$ for which $\{\varphi_{w_i} : \mathfrak{g}^{r_i} \to \mathfrak{g}\}$ are non-trivial, we have the following:*

(1) *If $m \geq Cd^3$ then $\varphi_{w_1} * \ldots * \varphi_{w_m}$ is flat.*
(2) *If $m \geq Cd^4$ then $\varphi_{w_1} * \ldots * \varphi_{w_m}$ is (FRS).*

THEOREM B (Theorem 4.10). *Let $\{w_i \in \mathcal{A}_{r_i}\}_{i\in\mathbb{N}}$ be a collection of algebra words of degree at most $d \in \mathbb{N}$. Then for any $n \in \mathbb{N}$ such that $\{\varphi_{w_i} : M_n^r \to M_n\}_{i\in\mathbb{N}}$ are generating, with M_n is the ring of $n \times n$-matrices over K, the assertions of Theorem A hold.*

Any word map $\varphi_w : \underline{G}^r \to \underline{G}$ can be degenerated to a map $\widetilde{\varphi}_w : \mathfrak{g}^r \to \mathfrak{g}$, called the symbol of φ_w, by taking the analogue of the lowest order non-vanishing term in the Taylor expansion of φ_w near $(e, \ldots, e) \in \underline{G}^r$. The Baker-Campbell-Hausdorff formula shows that $\widetilde{\varphi}_w$ is in fact a Lie algebra word map (Proposition 7.5). Theorem A, together with the fact that the (FRS) property behaves well under deformations (Corollary 2.12) give the following:

THEOREM C (Theorem 7.8). *Let $d \in \mathbb{N}$, let $\underline{G}$ be a connected semisimple K-group and let $\{\varphi_{w_i} : \underline{G}^{r_i} \to \underline{G}\}_{i=1}^m$ be a collection of non-trivial word maps with symbols $\widetilde{\varphi}_{w_i}$ of degree at most d. Then there exists $C > 0$ such that for any $m \geq C \cdot d^4$ the map $\varphi_{w_1} * \ldots * \varphi_{w_m}$ is (FRS) at $(e, \ldots, e)$.*

We now turn to Item (1) of Question 1.4. To quantitatively measure the singularities of the fibers of word maps, we introduce the following notions:

DEFINITION 1.7. Let $\varphi : X \to Y$ be a morphism between geometrically irreducible smooth K-varieties. Let $J_m(\varphi) : J_m(X) \to J_m(Y)$ be the corresponding m-th jet morphism (see Subsections 1.4.2 and 2.1) and let $X_{\varphi(x),\varphi}$ be the fiber of φ over $\varphi(x)$.

(1) φ is called *ε-flat* if for every $x \in X$ we have $\dim X_{\varphi(x),\varphi} \leq \dim X - \varepsilon \dim Y$.
(2) φ is called *ε-jet flat* (resp. *jet-flat*) if $J_m(\varphi)$ is ε-flat (resp. flat) for every $m \in \mathbb{N}$.

Note that a 1-flat morphism is just flat. It follows from **[Mus02]** that the notion of ε-jet flatness is closely related to the log canonical threshold of the pair $(X, X_{\varphi(x),\varphi})$, denoted $\operatorname{lct}(X, X_{\varphi(x),\varphi})$; the morphism φ is ε-jet flat if and only if $\operatorname{lct}(X, X_{\varphi(x),\varphi}) \geq \varepsilon \dim Y$ for all $x \in X$.

Let w be a word in $\mathcal{A}_r$ or $\mathcal{L}_r$. An easy calculation of dimensions shows that if φ_w^{*t} is jet-flat for some $t \in \mathbb{N}$, then φ_w is $\frac{1}{t}$-jet flat. Since (FRS) morphisms are in particular jet-flat (see Corollary 3.12), one can deduce the following from the proof of Theorems A and B:

THEOREM D (Theorem 4.11). *Let $w_1 \in \mathcal{A}_r$ and $w_2 \in \mathcal{L}_r$ be words of degree at most d.*

(1) *The map $\varphi_{w_1} : M_n^r \to M_n$ is $\frac{1}{25d^3}$-jet flat, for every $n \in \mathbb{N}$ such that φ_{w_1} is generating.*
(2) *The map $\varphi_{w_2} : \mathfrak{g}^r \to \mathfrak{g}$ is $\frac{1}{2 \cdot 10^5 d^3}$-jet flat, for every simple K-Lie algebra $\mathfrak{g}$ such that φ_{w_2} is non-trivial.*

We denote by $\ell(w)$ the length of a word $w \in F_r$ (e.g. $xyx^{-1}y^{-1}$ has length 4). Theorem D can be used to give lower bounds on the log canonical threshold of the fibers of group word maps.

THEOREM E (Theorem 7.9). *For any $1 \neq w \in F_r$, the map $\varphi_w : \mathrm{SL}_n^r \to \mathrm{SL}_n$ is $\frac{1}{10^{17}\ell(w)^9}$-jet flat. In particular, $\operatorname{lct}(\mathrm{SL}_n^r, \varphi_w^{-1}(g)) \geq \frac{\dim \mathrm{SL}_n}{10^{17}\ell(w)^9}$ for every $g \in \mathrm{SL}_n$.*

Theorem A and Theorem D provide an answer to Question 1.4 for algebra and Lie algebra words. Theorem E provides an answer to Part (1) of Question 1.4 for $\underline{G} = \mathrm{SL}_n$. Probabilistic applications of the above results are discussed in Subsection 1.2.3.

1.2.2. The commutator word map. Let $w_0 = xyx^{-1}y^{-1} \in F_r$ and $w = [X,Y] \in \mathcal{L}_r$ be the commutator word and Lie algebra word. In [**AA16**], it was shown that $\varphi_{w_0}^{*21}$ is (FRS) for all semisimple algebraic groups $\underline{G}$, and that φ_w^{*21} is (FRS) for all classical Lie algebras (see [**AA16**, Theorem VIII]). These bounds were improved to 2 for SL_n and $\mathfrak{sl}_n$ in [**Bud21**], and to 11 in the general case in [**Kap**]. Using a criterion of Mustaţă for rational singularities [**Mus01**, Theorem 0.1], and the methods developed in the course of the proof of Theorem A, we improve these bounds and significantly simplify the proof of [**AA16**]:

THEOREM F (Theorems 6.1 and 6.2). *Let $\underline{G}$ be a semisimple algebraic K-group, with Lie algebra $\mathfrak{g}$.*

(1) *$\varphi_{w_0}^{*2} : \underline{G}^4 \to \underline{G}$ and $\varphi_w^{*2} : \mathfrak{g}^4 \to \mathfrak{g}$ are flat.*
(2) *$\varphi_{w_0}^{*4} : \underline{G}^8 \to \underline{G}$ and $\varphi_w^{*4} : \mathfrak{g}^8 \to \mathfrak{g}$ are (FRS).*

REMARK 1.8. It is conjectured that $\varphi_{w_0}^{*2}$ and φ_w^{*2} are (FRS) for any semisimple algebraic group $\underline{G}$. For $\underline{G}$ not of type A, Theorem F gives the best known bounds.

The fibers of the commutator map φ_{w_0} are closely related, via a classical theorem of Frobenius, to the representation growth of compact p-adic groups and arithmetic groups (see [**AA16, AA18**] and [**Bud21**, Theorems 1.3-1.5]). Thus, Theorem F gives the following applications.

COROLLARY 1.9 (cf. [**AA16**, Theorem A], [**AA18**, Theorem B]). *Let $\underline{G}$ be a semisimple, simply connected algebraic $\mathbb{Q}$-group, and let G be either*

(1) *a compact open subgroup of $\underline{G}(\mathbb{Q}_p)$ for some prime p; or*
(2) *$\underline{G}(\mathbb{Z})$, provided that $\underline{G}$ has $\mathbb{Q}$-rank≥ 2.*

Then there exists a constant C such that, for any $N \in \mathbb{N}$,

$$\#\{N\text{-dimensional irreducible } \mathbb{C}\text{-representations of } G\} < C \cdot N^7.$$

We further have,

COROLLARY 1.10 (cf. [**AA16**, Theorem B]). *Let $\underline{G}$ be a complex semisimple algebraic group, and Σ_g be a closed orientable surface of genus $g \geq 4$. Then the deformation variety*

$$\mathrm{Hom}(\pi_1(\Sigma_g), \underline{G}(\mathbb{C})) \simeq (\varphi_{w_0}^{*g})^{-1}(e)$$

has rational singularities.

1.2.3. Algebraic interpretation of uniform mixing times and the p-adic probabilistic Waring type problem. Let $\mathcal{G}$ denote the collection of all algebraic group schemes $\underline{G}$, with $\underline{G}_{\mathbb{Q}}$ semisimple and simply connected. For a prime power $q = p^r$, we denote the unique unramified extension of $\mathbb{Q}_p$ of degree r by $\mathbb{Q}_q$, its ring of integers by $\mathbb{Z}_q$, and the maximal ideal of $\mathbb{Z}_q$ by $\mathfrak{m}_q$.

We study the family $\{\underline{G}(\mathbb{Z}_q)\}_{\underline{G}\in\mathcal{G},q}$ through the quotients $\{\underline{G}(\mathbb{Z}_q/\mathfrak{m}_q^k)\}_{\underline{G}\in\mathcal{G},q,k}$. Let $w \in F_r$ and let $\tau_{w,\underline{G},q,k} := (\varphi_w)_*(\pi_{\underline{G}^r(\mathbb{Z}_q/\mathfrak{m}_q^k)})$ be the pushforward of the uniform probability measure on $\underline{G}^r(\mathbb{Z}_q/\mathfrak{m}_q^k)$ with respect to $\varphi_w : \underline{G}^r \to \underline{G}$.

DEFINITION 1.11. For a real function f on a finite group G and $a \in \mathbb{R}_{\geq 1}$, set

$$\|f\|_a := (|G|^{a-1} \sum_{g\in G} |f(g)|^a)^{1/a} \text{ and } \|f\|_\infty = |G| \cdot \max_{g\in G} |f(g)|.$$

For a probability measure μ on a finite group G, set $S_\mu := \{g \in G : \mu(g) > 0\}$. If $e \in S_\mu$ and $\langle S_\mu \rangle = G$, then the random walk induced by μ converges to the uniform probability measure on G, i.e. $\lim\limits_{t\to\infty} \|\mu^{*t} - \pi_G\|_a = 0$ for any $a \in [1,\infty]$ (e.g. see [**LP17**, Theorem 4.9]). In particular, by [**JZ08**], this applies to each measure $\tau_{w,\underline{G},q,k}$.

Definition 1.12. The minimal $t_a \in \mathbb{N}$ such that $\|\mu^{*t_a} - \pi_G\|_a < \frac{1}{2}$ is called the L^a*-mixing time* of the random walk. At this point, rapid convergence occurs; for any $m \in \mathbb{N}$ we have $\|\mu^{*t_a m} - \pi_G\|_a < 2^{-m}$ (e.g. [**LP17**, Lemma 4.18]).

We would now like to define a uniform version of mixing time for families of word maps. Given a set of primes S, denote by $\mathcal{P}_S$ the set of all prime powers $p^k \in \mathbb{N}$ where $p \notin S$.

Definition 1.13. We say that a word $w \in F_r$

(1) is *almost* $\{\mathbb{F}_q\}$*-uniform in* L^a if $\forall \underline{G} \in \mathcal{G}$ there exists a finite set S of primes such that
$$\lim_{\mathcal{P}_S \ni q \to \infty} \left\|\tau_{w,\underline{G},q,1} - \pi_{\underline{G}(\mathbb{F}_q)}\right\|_a = 0.$$
(2) is *almost* $\{\mathbb{Z}_q\}$*-uniform in* L^a if $\forall \underline{G} \in \mathcal{G}$ there exists a finite set S of primes such that
$$\lim_{\mathcal{P}_S \ni q \to \infty} \left(\sup_{k\in\mathbb{N}} \left\|\tau_{w,\underline{G},q,k} - \pi_{\underline{G}(\mathbb{Z}_q/\mathfrak{m}_q^k)}\right\|_a\right) = 0.$$

Let $\mathcal{F}$ be $\{\mathbb{Z}_q\}$ or $\{\mathbb{F}_q\}$. We say that w has an *almost* $\mathcal{F}$*-uniform* L^a*-mixing time* $t_{a,\mathcal{F}}$ if it is minimal such that $w^{*t_{a,\mathcal{F}}}$ is almost $\mathcal{F}$-uniform in L^a.

In Chapter 8 we utilize the Lang-Weil bounds to obtain a number theoretic characterization of flatness (see Theorem 8.4). In a separate work [**CGH**], complementary to this paper, we showed that the (FRS) property of a $\mathbb{Q}$-morphism $\varphi : X \to Y$ between smooth $\mathbb{Q}$-varieties can be interpreted by certain uniform bounds on the fibers of $\varphi : X(\mathbb{Z}_q/\mathfrak{m}_q^k) \to Y(\mathbb{Z}_q/\mathfrak{m}_q^k)$ (see further discussion in Section 1.4.3 in the simpler case where $q = p$ is prime). This allows us to give an algebraic interpretation of the notion of almost uniformity in L^∞.

Theorem G (Corollary 9.9). *Let* $w \in F_r$.

(1) w *is almost* $\{\mathbb{F}_q\}$*-uniform in* L^∞ *if and only if* $(\varphi_w)_{\mathbb{Q}} : \underline{G}_{\mathbb{Q}}^r \to \underline{G}_{\mathbb{Q}}$ *is flat and has geometrically irreducible fibers for every* $\underline{G} \in \mathcal{G}$.
(2) w *is almost* $\{\mathbb{Z}_q\}$*-uniform in* L^∞ *if and only if* $(\varphi_w)_{\mathbb{Q}} : \underline{G}_{\mathbb{Q}}^r \to \underline{G}_{\mathbb{Q}}$ *is (FRS) and has geometrically irreducible fibers for every* $\underline{G} \in \mathcal{G}$.

In Section 8 we furthermore provide a number theoretic interpretation of ε-jet flatness:

Theorem H (Theorem 8.18). *Let* $\varphi : X \to Y$ *be a dominant morphism between finite type* $\mathbb{Z}$*-schemes* X *and* Y, *with* $X_{\mathbb{Q}}, Y_{\mathbb{Q}}$ *smooth and geometrically irreducible, and let* $0 < \varepsilon \le 1$. *Then the following are equivalent:*

(1) *For any* $0 < \varepsilon' < \varepsilon$ *there exists a finite set* S *of primes, such that for every* $q \in \mathcal{P}_S$, *every* $k \in \mathbb{N}$ *and every* $y \in Y(\mathbb{Z}_q/\mathfrak{m}_q^k)$,
$$\left|\varphi^{-1}(y)\right| < q^{k(\dim X_{\mathbb{Q}} - \varepsilon' \dim Y_{\mathbb{Q}})}.$$
(2) $\varphi_{\mathbb{Q}} : X_{\mathbb{Q}} \to Y_{\mathbb{Q}}$ *is* ε*-jet flat.*

Theorems E and H yield the following p-adic analogue of [**LS12**].

THEOREM I (Corollary 9.10). *Let $1 \neq w \in F_r$ be a word. Then for any $n \in \mathbb{N}$ there exists a finite set $S(n)$ of primes such that for every $q \in \mathcal{P}_{S(n)}$ and every $g \in \mathrm{SL}_n(\mathbb{Z}_q/\mathfrak{m}_q^k)$ we have*

$$\tau_{w,\mathrm{SL}_n,q,k}(g) \leq \left|\mathrm{SL}_n(\mathbb{Z}_q/\mathfrak{m}_q^k)\right|^{-\frac{1}{2\cdot 10^{17}\ell(w)^9}}.$$

For any $\underline{G} \in \mathcal{G}$ and a prime p, write $\underline{G}^1(\mathbb{Z}_p)$ for the first congruence subgroup of $\underline{G}(\mathbb{Z}_p)$ (i.e. the kernel of $\underline{G}(\mathbb{Z}_p) \to \underline{G}(\mathbb{F}_p)$), and let $\mu_{\underline{G}^1(\mathbb{Z}_p)}$ be the normalized Haar measure. Using Theorem C and an equivalent analytic criterion for the (FRS) property ([**AA16**, Theorem 3.4]), we can provide L^∞-bounds for the random walk on $\underline{G}^1(\mathbb{Z}_p)$ induced from w:

THEOREM J (Corollary 9.11). *Let $1 \neq w \in F_r$ be a word. Then there exists $C > 0$, such that for every $\underline{G} \in \mathcal{G}$, every prime $p > p_0(\underline{G})$, and every $t \geq C\ell(w)^4$, the pushforward measure $(\varphi_w^{*t})_*(\mu_{\underline{G}^1(\mathbb{Z}_p)}^{rt})$ on $\underline{G}^1(\mathbb{Z}_p)$ has continuous density.*

The following theorem provides an algebraic interpretation of the notion of almost uniformity in L^1. This is a strengthening of [**LST19**, Theorem 2].

THEOREM K (Corollary 9.9). *Let $w \in F_r$. The following are equivalent:*
(1) *w is almost $\{\mathbb{F}_q\}$-uniform in L^1.*
(2) *w is almost $\{\mathbb{Z}_q\}$-uniform in L^1.*
(3) *$(\varphi_w)_{\mathbb{Q}} : \underline{G}_{\mathbb{Q}}^r \to \underline{G}_{\mathbb{Q}}$ has a geometrically irreducible generic fiber, for every $\underline{G} \in \mathcal{G}$.*

In [**LST19**, Theorem 1] it was shown that any convolution $w_1 * w_2$ of two non-trivial words is almost $\{\mathbb{F}_q\}$-uniform in L^1. Theorem K allows us to push this result to compact p-adic groups:

THEOREM L (Corollary 9.12). *Let $w_1 \in F_{r_1}$ and $w_2 \in F_{r_2}$ be non-trivial words. Then $w = w_1 * w_2$ is almost $\{\mathbb{Z}_q\}$-uniform in L^1.*

It follows from [**GS09**] that the commutator word is almost $\{\mathbb{F}_q\}$-uniform in L^1. Theorems F, G and K give the following:

COROLLARY 1.14. *Let $w_0 = g_1 g_2 g_1^{-1} g_2^{-1} \in F_2$ be the commutator word.*
(1) *$w_0 * w_0 * w_0 * w_0$ is almost $\{\mathbb{Z}_q\}$-uniform in L^∞.*
(2) *w_0 is almost $\{\mathbb{Z}_q\}$-uniform in L^1.*

1.3. Further discussion of the main results

Questions 1.3 and 1.4 (excluding the L^1-Waring type problems) have been tackled only quite recently, and mainly in the case of finite simple groups. In [**LS12**] a bound of the form $|G|^{r-\varepsilon(w)}$ was given on the size of the fibers of $\varphi_w : G^r \to G$ for large enough finite simple groups. In [**LST19**], the probabilistic Waring problem for finite simple groups received thorough treatment. We say that $w \in F_r$ is *almost $\mathcal{G}_{\mathrm{fin}}$-uniform in L^a* if $\lim\limits_{G \in \mathcal{G}_{\mathrm{fin}}:|G| \to \infty} \left\|\tau_{w,G} - \pi_G\right\|_a = 0$. It was shown in [**LST19**], that any word $w \in F_r$ has an almost $\mathcal{G}_{\mathrm{fin}}$-uniform L^1-mixing time $t_{1,\mathrm{fin}}(w) \leq 2$ and an almost $\mathcal{G}_{\mathrm{fin}}$-uniform L^∞-mixing time $t_{\infty,\mathrm{fin}}(w) \leq C \cdot l^4$, for some explicit $C \gg 1$ (see [**LST19**, Theorems 1 and 4]). As a corollary, the authors deduced that for any

$w \in F_r$ and any semisimple algebraic group $\underline{G}$, the map $\varphi_w^{*t} : \underline{G}^r \to \underline{G}$ is flat for any $t \geq C \cdot l^4$ (see [**LST19**, Theorem 5(i)]).

One if the main obstacle in the compact p-adic probabilistic Waring problem, is that unlike in the case of finite simple groups, their representation theory is much more involved and far less understood. In particular, there is no developed theory of character bounds as in the case of finite simple groups (e.g. [**BLST18, GLT20, GLT**]), and the use of other representation theoretic techniques is limited as well.

Theorems D, E and I can be seen as an algebro-geometric and compact p-adic versions of [**LS12**]. Theorem A can be seen as a Lie algebra analogue of [**LST19**, Theorem 5(i)], which is richer in the sense that it also gives information on the singularities of the fibers. Several interesting questions arise:

QUESTION 1.15.

(1) *Can we find* $\alpha, C > 0$ *such that for every* $w \in F_r$ *and every simple algebraic group* $\underline{G}$*, the word map* $\varphi_w^{*C\ell(w)^{\alpha}}$ *is (FRS)?*
(2) *If such* α *exists, what is its optimal value? What is the optimal value of such* α *if we just demand that our word map becomes flat?*
(3) *What is the answer to (2) in the case of simple Lie algebras, where* $\ell(w)$ *is replaced by* $\deg(w)$*?*

Theorems A and C provide a great deal of evidence towards a positive answer to (1). In fact, in [**AA16**, Theorem IV] it was shown that the commutator map is (FRS) after t convolutions if and only if it is (FRS) at $(e, \ldots, e) \in \underline{G}^{2t}$ after t convolutions. This leads to the following question:

QUESTION 1.16. *Can we characterize or find nice families of words* $w \in F_r$*, whose worst singularities are concentrated at* $(e, \ldots, e)$*? That is, which words* $w \in F_r$ *satisfy the property that for every simple algebraic group* $\underline{G}$*,* φ_w^{*t} *is (FRS)/flat if and only if* φ_w^{*t} *is (FRS)/flat at* $(e, \ldots, e) \in \underline{G}^{rt}$*.*

EXAMPLE 1.17.

(1) Let $w = x^m$. Then $\varphi_w : \underline{G} \to \underline{G}$ is smooth at $(e, \ldots, e)$ for every simple algebraic group $\underline{G}$. On the other hand, by analyzing the fiber over $e \in \underline{G}$, one sees that φ_w is not even flat, and that in fact $\varphi_w^{*(m-1)}$ is not flat.
(2) Let $w_n := [[\ldots[[X, Y], Y], \ldots], Y] \in \mathcal{L}(X, Y)$ be the Engel word (where Y appears n times). Then $\varphi_{w,n}^{*(n-1)} : \mathfrak{g}^{2n-2} \to \mathfrak{g}$ is not flat.

We therefore see that not all words satisfy the condition of Question 1.16 and moreover, that a potentially optimal α for Items (2) and (3) of Question 1.15 is at least 1. There are certain low-rank phenomena for word maps (discussed in Sections 4 and 5), that is, when the length (resp. degree) of the word is large compared to the rank of $\underline{G}$ (resp. $\mathfrak{g}$). We therefore expect different answers for Items (2) and (3) of Question 1.15 when considering only groups or Lie algebras of high enough rank.

1.4. Methods and main ideas of the proofs

1.4.1. Singularities of word maps: Theorems A-E. The proof of Theorem A was inspired by [**AA16**], where the special case of the commutator word map was analyzed. We generalize some of the techniques in [**AA16**] as well as introduce new ones. The main principal employed is that many singularity properties, such as flatness and the (FRS) property, are preserved under smooth deformations. In other words, given a Lie algebra word map $\varphi_w : \mathfrak{g}^r \to \mathfrak{g}$, we may degenerate it to a map ψ_w which is easier to handle, and prove that ψ_w is flat or (FRS) instead. The idea is to apply in each time one of the following operations to a given morphism:

(1) Taking a self-convolution of our morphism;
(2) Applying a degeneration to our morphism,

such that:

(1) The use of Operation (1) will be as economical as possible. Explicitly, we would like the number of convolutions to be bounded by a polynomial in the degree of $w \in \mathcal{L}_r$, regardless of the simple Lie algebra under consideration.
(2) Every time we apply a degeneration, the resulting morphism should stay in a certain nice predefined class of morphisms, and in particular it should be generating.
(3) After finitely many steps as above we want to arrive at a morphism we can easily show is flat or (FRS).

In order to deal with the added complexities of arbitrary words, we developed a systematic approach to degenerations. We introduced three different kinds of degeneration techniques (averaging, monomialization, and level separation), each with a different purpose, as well as a general procedure to execute them (see Section 5.1). We further introduced degeneration techniques "in the direction of the jet" (Section 5.1.1), which served both as a complementary brute-force technique (see Item (5) below) and as a way to reduce questions about the (FRS) property to questions about flatness. A sketch of the key steps of the proof is as follows:

(1) In Section 4.4, we reduce to the case of self-convolutions of homogeneous word maps.
(2) Generalizing [**AA16**], we attach to each d-homogeneous word map $\varphi_w : \mathfrak{g}^r \to \mathfrak{g}$ a certain "combinatorial gadget" which we call a d-polyhypergraph, and is a graph whose hyperedges consist of multisets of d vertices and have types associated to them. Each hyperedge corresponds to a sum of certain monomials in φ_w, and the types correspond to a choice of coordinates on the target. We make sense of what it means for such an object to be flat or (FRS).
(3) We encode the two operations above combinatorially. The convolution operation corresponds to adding another copy of the polyhypergraph with hyperedges of identical types, and degeneration corresponds to eliminating certain hyperedges which are of the same type.
(4) In Sections 5.3 and 5.5, we deal with d-homogeneous word maps on high rank classical Lie algebras, and show that by applying an economical series of convolutions and degenerations to the corresponding d-polyhypergraph, one gets a hypergraph with a certain tame combinatorial structure (for example, a disjoint union of hypergraphs each consisting of a single hyperedge).
(5) In Section 5.2 we deal with the case of low rank simple Lie algebras (when $\mathrm{rk}(\mathfrak{g}) = O(d)$), and with the case of a tame combinatorial object (as in (4)). The case of a hypergraph consisting of a single hyperedge corresponds to a d-homogeneous polynomial map $\varphi : \mathbb{A}^n \to \mathbb{A}^1$. For such maps, one can give an effective lower bound on the log canonical threshold of the

fibers. Using a Thom-Sebastiani type result, we can deduce that $\varphi^{*(d+1)}$ is (FRS).

There are certain difficulties that we deal with. In general, a generating morphism might become non-generating after degeneration, so we need to choose our degenerations carefully; for example, $\psi_t(x,y) = (x, x+ty^2)$ is a degeneration of a dominant map $\psi_1(x,y) = (x, x+y^2)$ to a non-generating map $\psi_0(x,y) = (x,x)$. In addition, there are certain low rank phenomena in Lie algebra word maps that make the choice of degenerations especially difficult. Firstly, words of degree d can be trivial on Lie algebras of comparable rank (see Example 4.3). Moreover, certain degenerations can be proved to be effective only when the rank is large enough ($\sim d$). Since the degree d of the words could be arbitrary large, Step (5) is crucial.

The idea of the proof of Theorem E is to reduce to the case of matrix word maps, which is Theorem D(1). The main observation (Lemma 7.10) is that SL_{2n} is boundedly generated by matrices of the form

$$\left\{\begin{pmatrix} I_n & A \\ 0 & I_n \end{pmatrix} : A \in M_n\right\} \text{ and } \left\{\begin{pmatrix} I_n & 0 \\ B & I_n \end{pmatrix} : B \in M_n\right\},$$

and that $\begin{pmatrix} I_n & A \\ 0 & I_n \end{pmatrix}^{-1} = \begin{pmatrix} I_n & -A \\ 0 & I_n \end{pmatrix}$. This allows us to write φ_w in block matrix form, where each block is a polynomial of matrices (in M_n), which we know how to deal with. This is done in Section 7.

1.4.2. The commutator word map. Given a K-scheme

$$X = \mathrm{Spec} K[x_1, \ldots, x_n]/(f_1, \ldots, f_k)$$

and $m \in \mathbb{N}$, one may define its *m-th jet scheme* $J_m(X)$, which is the scheme having the following coordinate ring:

$$K[x_1, \ldots, x_n, x_1^{(1)}, \ldots, x_n^{(1)}, \ldots, x_1^{(m)}, \ldots, x_n^{(m)}]/(\{f_j^{(u)}\}_{j=1,u=1}^{k,m}),$$

where $f_i^{(u)}$ is the u-th formal derivative of f_i. Given a morphism $\varphi : X \to Y$ of affine K-schemes, it induces a morphism $J_m(\varphi) : J_m(X) \to J_m(Y)$ on the corresponding jet schemes called the *m-th jet morphism*, which is given by $J_m(\varphi) = (\varphi, \varphi^{(1)}, \ldots, \varphi^{(m)})$, with $\varphi^{(u)}$ again being the u-th formal derivative of φ (as a polynomial map). See Section 2.1 for a more general (and precise) definition. Using this language, Mustaţă introduced the following criterion for rational singularities:

THEOREM 1.18 ([**Mus01**]). *Let X be a geometrically irreducible, local complete intersection variety. Then the m-th jet scheme $J_m(X)$ of X is geometrically irreducible for all $m \in \mathbb{N}$ if and only if X has rational singularities.*

The relative formulation of this statement is as follows:

COROLLARY 1.19 (Corollary 3.12). *Let X and Y be smooth, geometrically irreducible K-varieties, and let $\varphi : X \to Y$ be a flat morphism with geometrically irreducible fibers. Then φ is (FRS) if and only if the m-th jet map $J_m(\varphi) : J_m(X) \to J_m(Y)$ is flat, with geometrically irreducible fibers for each $m \in \mathbb{N}$.*

We study the commutator map through its jets. By [**AA16**, Theorem IV], to prove Theorem F it is enough to consider the Lie algebra case, and to show that $\varphi_w^{*4} : \mathfrak{g}^8 \to \mathfrak{g}$ is (FRS) at $(0, \ldots, 0) \in \mathfrak{g}^8$, where $\varphi_w(X,Y) = [X,Y]$. Using Corollary 1.19, we may consider the m-th jet $J_m(\varphi_w) : J_m(\mathfrak{g}^2) \to J_m(\mathfrak{g})$ of the commutator map, and it is enough to show that each $J_m(\varphi_w)$ is flat at $(0,\dots,0)\in\mathfrak{g}^{2m}$, and that $J_m(\varphi_w^{-1}(0))$ is integral at $(0,\dots,0)$. Since the property of being flat and having locally integral fibers is preserved under deformations (Proposition 2.10), we may use degeneration methods as discussed in the previous subsection.

Under the identification $J_m(\mathfrak{g})\simeq\mathfrak{g}^{m+1}$, the map $J_m(\varphi_w)$ is of the form

$$(\varphi_w,\varphi_w^{(1)},\dots,\varphi_w^{(m)}):\mathfrak{g}^{2(m+1)}\to\mathfrak{g}^{m+1},$$

where each $\varphi_w^{(u)}$ is the u-th formal derivative of φ_w, and in particular a Lie algebra word map of degree 2. By applying a self-convolution, and a series of degenerations, the map $J_m(\varphi_w^{*2})$ is degenerated to a map comprised of commutators in disjoint variables,

$$\phi_m(Y_1,Y_2,\dots,Y_{2m-1},Y_{2m})=([Y_1,Y_2],[Y_3,Y_4],\dots,[Y_{2m-1},Y_{2m}]),$$

whose fibers are a product of fibers of commutator maps. Hence, it is enough to show that φ_w^{*2} is flat, with reduced, geometrically irreducible fibers, which is done in Theorem 6.2 using Fourier analysis.

It is worth noting that flatness of the commutator map is known in the group case ([**Li93**] and [**Sha09**, Theorem 2.8]). In the Lie algebra case, except of $\mathfrak{sl}_n$, where a stronger claim is proved in [**Bud21**], no other proof was known to the authors.

1.4.3. Number theoretic interpretation of the (FRS) and ε-jet flatness properties and applications. Let X be a finite type $\mathbb{Z}$-scheme. The Lang-Weil estimates [**LW54**] give the following for almost any prime p:

$$|X(\mathbb{F}_p)|=p^{\dim X_{\mathbb{Q}}}\left(C_{X,p}+O(p^{-1/2})\right),$$

where $C_{X,p}$ is the number of top dimensional irreducible components of $X_{\overline{\mathbb{F}}_p}$ which are defined over $\mathbb{F}_p$. In particular we see that $\left\{\frac{|X(\mathbb{F}_p)|}{p^{\dim X_{\mathbb{Q}}}}\right\}_{p\in\text{primes}}$ is uniformly bounded. Given a morphism $\varphi:X\to Y$ between finite type smooth schemes, the flatness of $\varphi_{\mathbb{Q}}$ can be shown to be equivalent to certain uniform bounds on the fibers of $\varphi:X(\mathbb{F}_q)\to Y(\mathbb{F}_q)$ over all finite fields (see Theorem 8.4).

For a finite ring A, set $h_X(A):=\frac{|X(A)|}{|A|^{\dim X_{\mathbb{Q}}}}$ if $X_{\mathbb{Q}}$ is non-empty, and $h_X(A)=0$ otherwise. Unlike the case of finite fields, when considering points over finite rings, the singularities of X come into play. For example, for $X=\mathrm{Spec}\mathbb{Z}[x]/(x^2)$, we have $h_X(\mathbb{Z}/p^{2k}\mathbb{Z})=p^k$, which is not uniformly bounded in neither p nor k. On the other end of the spectrum, if X is smooth, a generalization of Hensel's Lemma implies that $\left\{h_X(\mathbb{Z}/p^k\mathbb{Z})\right\}_{p,k}$ is uniformly bounded. The following natural question arises:

QUESTION 1.20. *Can we characterize finite type $\mathbb{Z}$-schemes X, such that*

$$\left\{h_X(\mathbb{Z}/p^k\mathbb{Z})\right\}_{p,k}$$

is uniformly bounded?

In [**AA18**], Aizenbud and Avni characterized this class of schemes in the case where $X_{\mathbb{Q}}$ is a local complete intersection. This was done relying on results of Mustaţă [**Mus01**] and Denef [**Den87**], and was improved in [**Gla19**].

THEOREM 1.21 (cf. [**AA18**, Theorem A], [**Gla19**, Theorem 1.4]). *Let X be a finite type $\mathbb{Z}$-scheme, where $X_{\mathbb{Q}}$ is equidimensional and a local complete intersection.*

(1) *$X_{\mathbb{Q}}$ is reduced and has rational singularities if and only if there exists $C > 0$ such that $h_X(\mathbb{Z}/p^k\mathbb{Z}) < C$, for every prime p, and every $k \in \mathbb{N}$.*
(2) *$X_{\mathbb{Q}}$ is geometrically irreducible, reduced and has rational singularities if and only if there exists $C > 0$ such that $\left|h_X(\mathbb{Z}/p^k\mathbb{Z}) - 1\right| < Cp^{-1/2}$, for every prime p, and every $k \in \mathbb{N}$.*

Notice that an (FRS) morphism $\varphi_{\mathbb{Q}}$ is in particular flat, so its fibers are local complete intersections (assuming the base and target are smooth). The condition of rational singularities puts each individual $\mathbb{Z}$-fiber of φ in the framework of Theorem 1.21. It is therefore expected that the (FRS) property of a morphism φ, should imply certain bounds on the size of the fibers of $\varphi : X(\mathbb{Z}/p^k\mathbb{Z}) \to Y(\mathbb{Z}/p^k\mathbb{Z})$. The fact that these bounds are uniform both in p, k and $y \in Y(\mathbb{Z}/p^k\mathbb{Z})$ is a rather deep result which we proved in a separate paper, joint with Raf Cluckers (see [**CGH**, Theorem A]). This is the key ingredient in the proof of Theorem G.

Similarly to the (FRS) property, Theorem H characterizes the ε-jet flatness property in terms of certain uniform bounds on the size of the fibers over finite rings of the form $\mathbb{Z}/p^k\mathbb{Z}$. The idea of the proof is to first reduce to the study of the $\mathbb{F}_p[t]/t^k$-fibers of φ, or equivalently, the fibers of $J_{k-1}(\varphi) : J_{k-1}(X)(\mathbb{F}_p) \to J_{k-1}(Y)(\mathbb{F}_p)$. The Lang-Weil bounds on the k-th jet maps $J_k(\varphi)$, connect between the ε-flatness of $J_{k-1}(\varphi)$ and the desired upper bound on the $\mathbb{F}_p$-fibers of $J_{k-1}(\varphi)$, for each fixed k. Since the family of morphisms $\{J_k(\varphi)\}_{k\in\mathbb{N}}$ is of unbounded complexity, a finer treatment is needed.

It can be shown that the function $g : Y(\mathbb{Z}_p) \times \mathbb{N} \to \mathbb{R}$ defined by $g(y,k) := \left|\varphi^{-1}(r_k(y))\right|$, where $r_k : Y(\mathbb{Z}_p) \to Y(\mathbb{Z}/p^k\mathbb{Z})$ is the reduction map, belongs to a certain class of functions with a well behaved structure and integration theory, called *motivic functions* (in the sense of [**CL08, CL10**], see Section 8.3 for more details). We then use results in motivic integration, to deal with the unbounded complexity of $\{J_k(\varphi)\}_{k\in\mathbb{N}}$ and finish the proof.

Most of the probabilistic applications follow from the algebro-geometric results (Theorems A-F) and the number theoretic interpretations of the flatness, ε-jet flatness and the (FRS) properties. To prove Theorem K we generalize [**LST19**, Theorem 2]. The main observation of [**LST19**] is that the property that a word $w \in F_r$ is almost $\{\mathbb{F}_q\}$-uniform in L^1 has a geometric characterization, namely, that the generic fiber of $\varphi_w : \underline{G}^r \to \underline{G}$ is geometrically irreducible for every simple algebraic group $\underline{G}$. We show that if this is the case, then the m-th jet morphism $J_m(\varphi_w) : J_m(\underline{G}^r) \to J_m(\underline{G})$ also has a geometrically irreducible generic fiber (Lemma 3.14), which in turn implies that $w \in F_r$ is almost $\{\mathbb{Z}_q\}$-uniform.

1.5. Conventions

- By $\mathbb{N}$ we mean the set $\{0, 1, 2, \dots\}$.
- We denote the complement of a set S by S^c.
- Unless explicitly stated otherwise, K is a field of characteristic 0 and F is a non-Archimedean local field of characteristic 0 whose ring of integers is $\mathcal{O}_F$.
- For any scheme X and an element $x \in X$ we denote by $\kappa(x)$ its function field.
- For a morphism $\varphi : X \to Y$ of schemes, we denote by $X_{y,\varphi}$ the scheme theoretic fiber $\mathrm{Spec}(\kappa(y)) \times_Y X$ of φ over $y \in Y$. We sometimes denote it by X_y if the map φ is understood.

- For a K-morphism $\varphi : X \to Y$ between K-varieties X and Y, we denote by X^{sm} (resp. X^{sing}) the smooth (resp. non-smooth) locus of X, and we denote by $X^{\mathrm{sm},\varphi}$ (resp. $X^{\mathrm{sing},\varphi}$) the smooth (resp. non-smooth) locus of φ in X.
- For an S-scheme X, we denote the base change with respect to $S' \to S$ by $X_{S'}$.

CHAPTER 2

Preliminaries

2.1. Basic properties of jet schemes

For a thorough discussion of jet schemes and their properties see [**CLNS18**, Chapter 3] and [**EM09**].

DEFINITION 2.1 (cf. [**CLNS18**, Section 3.2]). Let S be a scheme and let X be a scheme over S.

(1) For any $m \in \mathbb{N}$, we define the m-th jet scheme of X, denoted $J_m(X/S)$ as the S-scheme representing the functor

$$\mathcal{J}_m(X/S) : W \longmapsto \mathrm{Hom}_{S\text{-schemes}}(W \times_{\mathrm{Spec}\mathbb{Z}} \mathrm{Spec}(\mathbb{Z}[t]/(t^{m+1})), X),$$

where W is an S-scheme. When the scheme S is understood, we omit it from our notation.

(2) Given an S-morphism $\varphi : X \to Y$ between two S-schemes X and Y, and given an S-scheme W, the composition with φ yields a map

$$\mathcal{J}_m(X/S)(W) \to \mathcal{J}_m(Y/S)(W),$$

which corresponds to a morphism

$$J_m(\varphi) : J_m(X/S) \to J_m(Y/S),$$

and is called the *m-th jet* of φ.

From now on we restrict ourselves to finite type K-schemes (with $S = \mathrm{Spec}K$).

DEFINITION 2.2.

(1) Let X be a finite type K-scheme. For any $m \geq n \in \mathbb{N}$ and any finite type K-algebra A, we have a natural map $A[t]/(t^{m+1}) \to A[t]/(t^{n+1})$ which induces, via $\mathcal{J}_m(X)(\mathrm{Spec}A) \to \mathcal{J}_n(X)(\mathrm{Spec}A)$, a natural morphism $\pi_n^m : J_m(X) \to J_n(X)$. The maps $\{\pi_n^m\}_{m\geq n}$ are called *truncation maps.* Note that any K-morphism $\varphi : X \to Y$ gives rise to a collection of morphisms $\{J_m(\varphi) : J_m(X) \to J_m(Y)\}_{m\in\mathbb{N}}$ between the corresponding m-th jet schemes, which commute with $\{\pi_n^m\}_{m\geq n}$.

(2) The natural map $A \to A[t]/(t^{m+1})$ induces a *zero section* $s_m : X \hookrightarrow J_m(X)$.

Let $X \subseteq \mathbb{A}^n$ be an affine K-scheme whose coordinate ring is

$$K[x_1, \ldots, x_n]/(f_1, \ldots, f_k).$$

Then the coordinate ring of $J_m(X)$ is

$$K[x_1, \ldots, x_n, x_1^{(1)}, \ldots, x_n^{(1)}, \ldots, x_1^{(m)}, \ldots, x_n^{(m)}]/(\{f_j^{(u)}\}_{j=1,u=1}^{k,m}),$$

where $f_i^{(u)}$ is the u-th formal derivative of f_i. Here, the zero section is

$$s_m(x_1,\dots,x_n)=(x_1,\dots,x_n,0,\dots,0)\in J_m(X).$$

EXAMPLE 2.3.

(1) $J_m(\mathbb{A}^n)\simeq\mathbb{A}^{n(m+1)}$.

(2) Let $X=\operatorname{Spec}(K[x]/(x^n))$. Then

$$J_2(X)=\operatorname{Spec}\left(K[x,x^{(1)},x^{(2)}]/(x^n,nx^{n-1}x^{(1)},n(n-1)x^{n-2}(x^{(1)})^2+nx^{n-1}x^{(2)})\right).$$

(3) For any smooth K-scheme X, the first jet scheme $J_1(X)$ can be identified with the tangent bundle TX.

Let $\varphi:\mathbb{A}^{n_1}\to\mathbb{A}^{n_2}$ be a morphism between affine spaces. Then $J_m(\varphi):\mathbb{A}^{n_1(m+1)}\to\mathbb{A}^{n_2(m+1)}$ is given by formally deriving φ, $J_m(\varphi)=(\varphi,\varphi^{(1)},\dots,\varphi^{(m)})$. By functoriality, the m-th jet $J_m(\varphi)$ of a morphism $\varphi:X\to Y$ of affine K-schemes, is given by the formal derivative of φ.

EXAMPLE 2.4. Let $\phi:\mathfrak{gl}_n^2\to\mathfrak{gl}_n$ be the commutator map $\phi(X,Y)=[X,Y]$. Then

$$\begin{aligned}\phi_1(X,Y,X^{(1)},Y^{(1)})&=\left([X,Y],XY^{(1)}+X^{(1)}Y-YX^{(1)}-Y^{(1)}X\right)\\&=\left([X,Y],[X,Y^{(1)}]+[X^{(1)},Y]\right),\end{aligned}$$

where $X^{(1)}$ is the formal derivative of X (by deriving each entry). We will use these kind of computations in Chapters 4, 5 and 6.

Here are two useful facts:

LEMMA 2.5.

(1) *If X is a smooth K-variety, then $J_m(X)$ is smooth of dimension $(m+1)\dim X$ for any $m\geq 1$, and all the truncation maps $\pi_n^m:J_m(X)\to J_n(X)$ are locally trivial fibrations with fiber $\mathbb{A}^{\dim X(m-n)}$.*

(2) *If $\varphi:X\to Y$ is an étale morphism then the commutative diagram*

$$\begin{array}{ccc}J_m(X)&\xrightarrow{J_m(\varphi)}&J_m(Y)\\ \downarrow\pi_0^m&&\downarrow\pi_0^m\\ X&\xrightarrow{\varphi}&Y\end{array}$$

is a fibered diagram (and in particular $J_m(\varphi)$ is étale).

2.2. Log canonical threshold

Let X be a smooth K-variety and let $Z\subseteq X$ be a closed subscheme. A *log resolution* $\pi:Y\to X$ of (X,Z) is a proper, birational morphism from a smooth variety Y, such that $D:=\pi^{-1}(Z)$ is a Cartier divisor and $D+K_{Y/X}$ is a divisor with simple normal crossings, where $K_{Y/X}$ is the relative canonical divisor defined by the determinant of the Jacobian of π (see [**Mus12**, Section 1.1]). Such a resolution exists by Hironaka [**Hir64**]. In particular $K_{Y/X}$ and D can be written as $K_{Y/X}=\sum_{i=1}^{N}k_iE_i$ and $D=\sum_{i=1}^{N}a_iE_i$, where the E_i are distinct smooth irreducible divisors.

The *log canonical threshold* of the pair (X,Z) is defined as

$$\operatorname{lct}(X,Z):=\min_{x\in X}\operatorname{lct}_x(X,Z),$$

where

$$\mathrm{lct}_x(X, Z) := \min_{i : x \in \pi(E_i)} \frac{k_i + 1}{a_i}.$$

By convention, if $x \notin Z$, we set $\mathrm{lct}_x(X, Z) = \infty$. For X affine and smooth, with coordinate ring $K[X]$, and an ideal $\mathfrak{a} \subseteq K[X]$, we write $\mathrm{lct}(\mathfrak{a}) := \mathrm{lct}(X, V(\mathfrak{a}))$, where $V(\mathfrak{a})$ is the vanishing set of $\mathfrak{a}$. In general, lower values of $\mathrm{lct}(X, Z)$ correspond to "wilder" singularities of the pair (X, Z). If $\mathrm{lct}(X, Z) \geq 1$ then the pair (X, Z) is *log canonical.* If Z is a smooth subscheme, then $\mathrm{lct}(X, Z) = \dim X - \dim Z$, which is the largest possible value.

Mustaţă showed that the log canonical threshold can be characterized in terms of the growth rate of the dimensions of the jet schemes of Z:

THEOREM 2.6 ([**Mus02**, Corollary 0.2], [**CLNS18**, Corollary 7.2.4.2]). *Let X be a smooth, geometrically irreducible K-variety, and let $Z \subsetneq X$ be a closed subscheme. Then*

$$\mathrm{lct}(X, Z) = \dim X - \sup_{m \geq 0} \frac{\dim J_m(Z)}{m + 1}.$$

Note that $\mathrm{lct}(X, Z)$ does not depend on the ambient space X (only on $\dim X$) or the embedding of Z in X.

EXAMPLE 2.7. Let $\varphi_i(x) = x_i^{n_i}$ for $1 \leq i \leq k$. Then $\mathrm{lct}(\mathbb{A}^1_{\mathbb{C}}, \varphi_i^{-1}(0)) = \mathrm{lct}_0(\mathbb{A}^1_{\mathbb{C}}, \varphi_i^{-1}(0)) = \frac{1}{n_i}$. By a Thom-Sebastiani type result [**MSS18**, Corollary 1] we have

$$\mathrm{lct}(\mathbb{A}^k_{\mathbb{C}}, (\varphi_1 * \cdots * \varphi_k)^{-1}(0)) = \min\{1, \sum_{i=1}^{k} \mathrm{lct}(\mathbb{A}^1_{\mathbb{C}}, \varphi_i^{-1}(0))\} = \min\{1, \sum_{i=1}^{k} \frac{1}{n_i}\}.$$

We therefore see that the convolution operation improves the log canonical threshold.

The following properties of the log canonical threshold will be useful for us.

FACT 2.8 (see [**Mus12**, Section 1]). *Let X be a smooth, geometrically irreducible, affine K-variety.*

(1) *Let $0 \neq \mathfrak{a}_1 \subseteq \mathfrak{a}_2$ be two ideals in $K[X]$. Then $\mathrm{lct}_x(\mathfrak{a}_1) \leq \mathrm{lct}_x(\mathfrak{a}_2)$ for any $x \in X$.*
(2) *Let $\mathfrak{a} \subseteq K[X]$ be an ideal. Then for any $x \in X$ we have $\mathrm{lct}_x(\mathfrak{a}) \geq \frac{1}{\mathrm{ord}_x(\mathfrak{a})}$, where*

$$\mathrm{ord}_x(\mathfrak{a}) = \min_{f \in \mathfrak{a}} \mathrm{ord}_x(f),$$

and $\mathrm{ord}_x(f)$ is the minimal $|\alpha|$ such that $\frac{\partial^\alpha f}{\partial x^\alpha}(x) \neq 0$.

2.3. A method of degeneration

DEFINITION 2.9. Let X and S be K-schemes. An *S-deformation* of X is a flat K-morphism $\pi : \widetilde{X} \to S$ such that $X_s \simeq X$ for some $s \in S$. Assume we are given a deformation $\pi : \widetilde{X} \to S$ such that $\widetilde{X}_s \simeq X$ for any s in a punctured neighborhood U of $s_0 \in S$. In this situation we say that $\widetilde{X}_{s_0}$ is an *S-degeneration* of X.

PROPOSITION 2.10. *Let S be a K-variety, let X and Y be smooth S-varieties, with structure maps $\pi_X : X \to S$ and $\pi_Y : Y \to S$, and let $\varphi : X \to Y$ be an S-morphism. Let P be one of the following properties of morphisms:*

(1) *Flatness.*
(2) *Flatness with reduced fibers.*
(3) *Flatness with locally integral (i.e. locally irreducible and reduced) fibers.*
(4) *Flatness with normal fibers.*
(5) *(FRS).*
(6) *Smoothness.*

Then the set of points $x \in X$ for which the fiber map $\varphi_{\pi_X(x)} : X_{\pi_X(x)} \longrightarrow Y_{\pi_X(x)}$ is P at x is open.

PROOF. Let $x \in X$ and $s_x := \pi_X(x)$. Assume that $\varphi_{s_x} : X_{s_x} \to Y_{s_x}$ is P at x. Since X and Y are S-smooth, by miracle flatness it follows for any $x' \in X$, that $\varphi_{s_{x'}}$ is flat at x' if and only if φ is flat at x'. This implies (1). By restricting to an open neighborhood, we may assume that φ is flat. Note that $(X_{s_x})_{\varphi_{s_x}(x),\varphi_{s_x}} \simeq X_{\varphi(x),\varphi}$. Now, since φ is flat it follows by [**Gro67**, Theorems 12.1.1, 12.1.6] and by [**Elk78**] that the set

$$\{x \in X : \varphi \text{ is } P \text{ at } x\} = \{x \in X : \varphi_{s_x} \text{ is } P \text{ at } x\}$$

is open. □

NOTATION 2.11. We denote by $\mathcal{P}$ the collection of singularity properties P appearing in Proposition 2.10.

Given an S-degeneration $\widetilde{\varphi} : \widetilde{X} \to \widetilde{Y}$ of a morphism $\varphi : X \to Y$ to a morphism $\varphi_0 := \widetilde{\varphi}_{s_0} : \widetilde{X}_{s_0} \to \widetilde{Y}_{s_0}$, and a property $P \in \mathcal{P}$, one can obtain information on the P-locus of φ using the P-locus of φ_0. This is manifested in the following corollary:

COROLLARY 2.12 (cf. [**AA16**, Corollary 2.3]). *Let $\widetilde{X}$ and $\widetilde{Y}$ be smooth $\mathbb{A}^1$-varieties with a $\mathbb{G}_m$ action which is compatible with the action of $\mathbb{G}_m$ on $\mathbb{A}^1$. Let $\widetilde{\varphi} : \widetilde{X} \to \widetilde{Y}$ be a $\mathbb{G}_m$-equivariant $\mathbb{A}^1$-morphism, let $s : \mathbb{A}^1 \to \widetilde{X}$ be a $\mathbb{G}_m$-equivariant section of the structure map $\widetilde{X} \to \mathbb{A}^1$ and let $P \in \mathcal{P}$. Assume that $\widetilde{\varphi}_0$ is P at $s(0)$. Then $\widetilde{\varphi}_1$ is P at $s(1)$.*

Let X be an algebraic K-variety and let $x \in X(\overline{K})$. The local ring $\mathcal{O}_{X,x}$ has natural descending filtration $\mathfrak{m}^i_{X,x}$, which induces a graded ring

$$\operatorname{gr}(\mathcal{O}_{X,x}) = \bigoplus_{i \geq 0} \mathfrak{m}^i_{X,x}/\mathfrak{m}^{i+1}_{X,x}.$$

We define the *geometric tangent cone* $C_x(X)$ of X at x to be $\operatorname{Spec}(\operatorname{gr}(\mathcal{O}_{X,x})_{\mathrm{red}})$, where $\operatorname{gr}(\mathcal{O}_{X,x})_{\mathrm{red}}$ is the quotient of $\operatorname{gr}(\mathcal{O}_{X,x})$ by its nilradical. Any morphism $\varphi : X \to Y$ induces a map $\varphi^* : \mathcal{O}_{Y,\varphi(x)} \to \mathcal{O}_{X,x}$, which yields a morphism of graded rings $\operatorname{gr}(\varphi^*) : \operatorname{gr}(\mathcal{O}_{Y,\varphi(x)}) \to \operatorname{gr}(\mathcal{O}_{X,x})$ and therefore induces a map $\operatorname{gr}(\varphi) : C_x X \to C_{\varphi(x)} Y$ between the tangent cones.

Since the map $\operatorname{gr}(\varphi)$ may be trivial, we would like to consider a slightly more general construction. Let $l \in \mathbb{N}$ satisfy $\varphi^*(\mathfrak{m}_{Y,\varphi(x)}) \subseteq \mathfrak{m}^l_{X,x}$, and define the following filtrations on $\mathcal{O}_{X,x}$ and $\mathcal{O}_{Y,\varphi(x)}$. For $\mathcal{O}_{X,x}$ we set $F^i\mathcal{O}_{X,x} = \mathfrak{m}^i_{X,x}$ for $i > 0$ and $F^0\mathcal{O}_{X,x} = \mathcal{O}_{X,x}$. For $\mathcal{O}_{Y,\varphi(x)}$ we define $F^i\mathcal{O}_{Y,\varphi(x)} = \mathfrak{m}^{\lfloor i/l \rfloor}_{Y,\varphi(x)}$ for $i > 0$ and $F^0\mathcal{O}_{Y,\varphi(x)} = \mathcal{O}_{Y,\varphi(x)}$. Our choice of l guarantees that the map φ^* is filtration-preserving, so φ induces a map on the tangent cones, denoted by

$$D^{l-1}_x := \operatorname{gr}_F(\varphi) : C_x X \to C_{\varphi(x)} Y.$$

By an appropriate choice of l we may guaranty that D_x^{l-1} is non-zero. In [**AA16**, Section 2.1.1] it is shown that any morphism $\varphi : X \to Y$ can be degenerated to the morphism $D_x^l\varphi : C_xX \to C_{\varphi(x)}Y$ for a suitable l. Corollary 2.12 gives the following:

COROLLARY 2.13 (cf. [**AA16**, Corollary 2.14], the linearization method). *Let $\varphi : X \to Y$ be a morphism of affine varieties, let $x \in X(\overline{K})$, and let $P \in \mathcal{P}$. Suppose that $D_x^l\varphi$ is P at x for some l for which $D_x^l\varphi$ is defined. Then φ is P at x.*

REMARK 2.14. Let $\varphi : \mathbb{A}^n \to \mathbb{A}^m$ be a morphism and assume that all derivatives at $(a_1, \dots, a_n) \in K^n$ of φ of order $\leq l$ vanish. Then $D_x^l\varphi$ is just the Taylor expansion of order $(l+1)$ at $(a_1, \dots, a_n)$ of the polynomial map φ.

The linearization method allows us to reduce the problem of showing that a morphism satisfies a certain singularity property $P \in \mathcal{P}$ to showing that some map between affine spaces satisfies P. We now treat the latter; for any $\omega = (\omega_1, \dots, \omega_n) \in \mathbb{Z}^n$ (referred to as a *weight*), we define the *ω-degree* of a monomial $x_1^{a_1} \cdot \dots \cdot x_n^{a_n} \in K[x_1, \dots, x_n]$ to be

$$\deg_\omega(x_1^{a_1} \cdot \dots \cdot x_n^{a_n}) := \sum_{i=1}^n a_i\omega_i.$$

For any polynomial $f \in K[x_1, \dots, x_n]$, we define the *symbol* of f, denoted $\sigma_\omega(f)$, to be the sum of the monomials of f with the lowest ω-degree. Notice that the degree induced from the weight $\omega = (1, \dots, 1)$ corresponds to the standard degree of monomials

$$\deg(x_1^{a_1} \cdot \dots \cdot x_n^{a_n}) = \sum_{i=1}^n a_i.$$

By choosing appropriate filtrations, and applying Corollary 2.12, one can obtain the following:

COROLLARY 2.15 (cf. [**AA16**, Corollary 2.17], the elimination method). *Let $\varphi = (\varphi_1, \dots, \varphi_m) : \mathbb{A}^n \to \mathbb{A}^m$ be a morphism, $\omega \in \mathbb{Z}^n$ be a weight, $P \in \mathcal{P}$ and let $\sigma_\omega(\varphi) = (\sigma_\omega(\varphi_1), \dots, \sigma_\omega(\varphi_m))$ be the symbol of φ. If $\sigma_\omega(\varphi)$ is P at $(0, \dots, 0)$, then so is φ.*

The above degeneration methods can be applied to jet morphisms. Given a K-variety X, there is a natural $\mathbb{G}_m$-action $\phi : \mathbb{G}_m \times J_m(X) \to J_m(X)$ defined on the level of A-valued points for any commutative ring A; for any $a \in A^\times$ and $f : \mathrm{Spec}A[t]/(t^{m+1}) \to X$ we set

$$\phi(a, f) = f \circ g_a,$$

where $g_a : \mathrm{Spec}A[t]/(t^{m+1}) \to \mathrm{Spec}A[t]/(t^{m+1})$ is induced from $t \mapsto at$.

Given a morphism $\varphi : X \to Y$, the map $J_m(\varphi) : J_m(X) \to J_m(Y)$ is $\mathbb{G}_m$-equivariant.

EXAMPLE 2.16. Let $X = \mathrm{Spec}\,(\mathbb{C}[x_1, \dots, x_n]/(f_1, \dots, f_k))$. Then any point $b \in J_m(X)(\mathbb{C})$ is of the form $b = (b_1, \dots, b_n, \dots, b_1^{(m)}, \dots, b_n^{(m)})$, where $f_j^{(u)}(b) = 0$ for $1 \leq j \leq k$ and $0 \leq u \leq m$. The above $\mathbb{G}_m$-action is then given by

$$s.b = (b_1, \dots, b_n, sb_1^{(1)}, \dots, sb_n^{(1)} \dots, s^m b_1^{(m)}, \dots, s^m b_n^{(m)}).$$

Note that $f_j^{(u)}(s.b) = s^u f_j^{(u)}(b) = 0$ as expected.

COROLLARY 2.17. *Let $\varphi : X \to Y$ be a morphism between smooth K-varieties and let $P \in \mathcal{P}$. Assume that $J_m(\varphi) : J_m(X) \to J_m(Y)$ is P at $(x, 0) \in J_m(X)$ for every $x \in X$. Then $J_m(\varphi)$ is P.*

PROOF. It is enough to prove the claim for X and Y affine. Set $\widetilde{J_m}(X) = J_m(X) \times \mathbb{A}^1$, $\widetilde{J_m}(Y) = J_m(Y) \times \mathbb{A}^1$ and $\widetilde{J_m}(\varphi) : \widetilde{J_m}(X) \to \widetilde{J_m}(Y)$ by

$$\widetilde{J_m}(\varphi)\left((x, x^{(1)}, \dots, x^{(m)}), t\right) = \left(J_m(\varphi)(x, x^{(1)}, \dots, x^{(m)}), t\right).$$

Note that $\widetilde{J_m}(\varphi)$ is $\mathbb{G}_m$-equivariant. Taking the $\mathbb{G}_m$-equivariant section $s(t) = (x, tx^{(1)}, \dots, t^m x^{(m)})$, and using the assumption that $J_m(\varphi)$ is P at $s(0) = (x, 0)$, by Corollary 2.12 we get that $J_m(\varphi)$ is P at $s(1) = (x, x^{(1)}, \dots, x^{(m)})$. □

CHAPTER 3

Some properties of the algebraic convolution operation

In this section we introduce various singularity properties and discuss their behavior with respect to the algebraic convolution operation. We first recall some results from [**GH19**] and [**GH21**] about morphisms from a smooth K-scheme to a K-algebraic group. Throughout this section we assume that K is a field of characteristic zero.

3.1. The algebraic convolution operation: The case of general morphisms

DEFINITION 3.1. Let $\varphi : X \to Y$ be a morphism between K-schemes.

(1) φ is called *normal at* $x \in X$ if φ is flat at x and the fiber $X_{\varphi(x)}$ is geometrically normal at x. φ is called *normal* if it is normal at every $x \in X$.
(2) φ is called *(FGI)* if it is flat with geometrically irreducible fibers.

REMARK 3.2. Since by convention the empty scheme is reducible, an (FGI) morphism is in particular surjective.

In [**GH19**] and [**GH21**], we have shown that the convolution operation improves various singularity properties of morphisms in the following sense:

PROPOSITION 3.3 ([**GH21**, Proposition 3.1, Lemma 3.4]). *Let X and Y be smooth geometrically irreducible K-varieties, $\underline{G}$ be an algebraic K-group and let S be any of the following properties of morphisms:*

(1) *Dominance.*
(2) *Flatness (or flatness with reduced fibers).*
(3) *Normality.*
(4) *Smoothness.*
(5) *(FGI).*
(6) *(FRS).*

Suppose $\varphi : X \to \underline{G}$ has property S and $\psi : Y \to \underline{G}$ is any morphism. Then the maps $\varphi * \psi : X \times Y \to \underline{G}$ and $\psi * \varphi : Y \times X \to \underline{G}$ have the property S.

By taking enough self-convolutions of a given dominant morphism, one can obtain increasingly better singularity properties, and eventually obtain an (FRS) morphism.

PROPOSITION 3.4 (cf. [**GH21**, Theorem B]). *Let $m \in \mathbb{N}$, let $X_1, \ldots, X_m$ be geometrically irreducible smooth K-varieties, $\underline{G}$ be a connected algebraic K-group and let $\{\varphi_i : X_i \to \underline{G}\}_{i=1}^{m}$ be a collection of dominant morphisms.*

(1) *For any* $1 \leq i, j \leq m$ *the morphism* $\varphi_i * \varphi_j$ *is surjective.*
(2) *If* $m \geq \dim \underline{G}$ *then* $\varphi_1 * \cdots * \varphi_m$ *is flat.*
(3) *If* $m \geq \dim \underline{G} + 1$ *then* $\varphi_1 * \cdots * \varphi_m$ *is flat with reduced fibers.*
(4) *If* $m \geq \dim \underline{G} + k$ *then* $\varphi_1 * \cdots * \varphi_m$ *is flat with normal fibers which are regular in codimension* $k - 1$.
(5) *The bounds in* (2) – (4) *are tight (e.g. consider* $\varphi(x_1, \ldots, x_m) = (x_1^2, (x_1x_2)^2, (x_1x_3)^2, (x_1x_m)^2)$*, see* [**GH21**, Proposition 3.6]*).*

THEOREM 3.5 ([**GH21**, Theorem A]). *Let X be a geometrically irreducible smooth K-variety, $\underline{G}$ be a connected K-algebraic group and let $\varphi : X \to \underline{G}$ be a dominant morphism. Then there exists $N \in \mathbb{N}$ such that for any $t > N$, the t-th convolution power φ^{*t} is (FRS).*

For the next sections we need several more auxiliary results.

PROPOSITION 3.6. *Let $\{X_i\}_{i\in\mathbb{N}}$ be a collection of geometrically irreducible smooth K-varieties, let $\underline{G}$ be a connected algebraic K-group, and let $\{\varphi_i : X_i \to \underline{G}\}_{i\in\mathbb{N}}$ be a collection of dominant morphisms. Assume that φ_1 is flat. Then*
(1) $\varphi_1 * \varphi_2$ *is flat with reduced fibers.*
(2) *If* $m \geq 3$ *then* $\varphi_1 * \cdots * \varphi_m$ *is flat with normal fibers which are regular in codimension* $m - 2$.

PROOF. Let $X = \prod_{i=1}^{k} X_i$. Notice that by Proposition 3.3, $\varphi := \varphi_1 * \varphi_2 * \cdots * \varphi_k : X \to \underline{G}$ is a flat morphism between smooth schemes, and therefore its fibers are local complete intersections and in particular Cohen-Macaulay. By Serre's criterion for normality and reducedness [**Gro67**, Proposition 5.8.5, Theorem 5.8.6] it is enough to show that for any $k \geq 2$ the fibers are regular in codimension $k - 2$.

Since φ is flat, $x = (x_1, \ldots, x_k)$ is a smooth point of the fiber $X_{\varphi(x),\varphi}$ if and only if φ is smooth at x. For each $1 \leq i \leq k$ denote by $\underline{G}^{\mathrm{sing},\varphi_i}$ the set of $g \in \underline{G}$ with $(X_i)_{g,\varphi_i}$ non-smooth. By generic smoothness, $\dim \varphi_i^{-1}(\underline{G}^{\mathrm{sing},\varphi_i}) \leq \dim X_i - 1$. Note that by Proposition 3.3, φ is non-smooth at x only if φ_i is non-smooth at x_i for each i. Thus, set theoretically, we have the following:

$$(\varphi^{-1}(g))^{\mathrm{sing}} \subseteq \bigcup_{(g_2,\ldots,g_k)\in\underline{G}^{\mathrm{sing},\varphi_2}\times\ldots\times\underline{G}^{\mathrm{sing},\varphi_k}} \varphi_1^{-1}(gg_k^{-1}\cdots g_2^{-1}) \times \varphi_2^{-1}(g_2) \times \cdots \times \varphi_k^{-1}(g_k).$$

In particular,

$$\dim(X_{g,\varphi}^{\mathrm{sing}}) \leq \dim X_1 - \dim \underline{G} + \sum_{i=2}^{k} \dim \varphi_i^{-1}(\underline{G}^{\mathrm{sing},\varphi_i})$$
$$\leq \sum_{i=1}^{k} \dim X_i - \dim \underline{G} - (k-1) = \dim X_{g,\varphi} - (k-1).$$

□

DEFINITION 3.7. Let $\varphi : X \to \underline{G}$ be a morphism between a geometrically irreducible K-variety X and a connected algebraic K-group $\underline{G}$. Then φ is called *generating*, if for any algebraic $\overline{K}$-subgroup $\underline{H} \lneq \underline{G}_{\overline{K}}$, and any $g \in \underline{G}(\overline{K})$ we have $\varphi(X) \not\subseteq g\underline{H}$.

LEMMA 3.8 (cf. [**GH19**, Proposition 3.5]). *Let X be a geometrically irreducible K-variety, $\underline{G}$ be a connected algebraic K-group and $\varphi : X \to \underline{G}$ a generating morphism whose image contains $e \in \underline{G}$. Then $\varphi^{*\dim\underline{G}}$ is dominant, and thus $\varphi^{*(\dim\underline{G})^2}$ is flat.*

PROOF. The proof is the same as in [**GH19**, Proposition 3.5]. It is only left to observe that $\overline{\varphi(X)^n}$ must stabilize for some $n \leq \dim\underline{G}$, and then using [**GH19**, Proposition 3.5] we deduce that $\overline{\varphi(X)^n} = \underline{G}$, so that φ^{*n} is dominant. $\square$

The following will be useful in Section 5.4.

LEMMA 3.9. *Let X and Y be smooth geometrically irreducible K-varieties, $\underline{G}$ a connected K-algebraic group, and $k \in \mathbb{N}$. Let $\psi : X \to \underline{G}$ and $\varphi : Y \to \underline{G}$ be dominant morphisms such that $\dim Y_{g,\varphi} \leq \dim Y - \dim\underline{G} + k$, for each $g \in \underline{G}$. Then $\varphi * \psi^{*k}$ is flat.*

PROOF. By induction, it suffices to show that $\varphi * \psi$ satisfies

$$\dim\left((\varphi * \psi)^{-1}(g)\right) \leq \dim X + \dim Y - \dim\underline{G} + k - 1.$$

Set $S_l := \{g \in \underline{G} : \dim X_{g,\psi} = \dim X - \dim\underline{G} + l\}$, and notice that set theoretically

$$(\varphi * \psi)^{-1}(g) = \bigcup_{l=0}^{\dim\underline{G}} \bigcup_{g_0 \in S_l} \varphi^{-1}(gg_0^{-1}) \times \psi^{-1}(g_0), \tag{3.1}$$

for any $g \in \varphi * \psi(X)$. Note that $\underline{G} = \bigcup_{l=0}^{\dim\underline{G}} S_l$, $X = \bigcup_{l=0}^{\dim\underline{G}} \psi^{-1}(S_l)$ and that $S_0 \subseteq \underline{G}$ and $X_0 := \psi^{-1}(S_0) \subseteq X$ are open subsets, by generic flatness of ψ. Hence for every $g \in G$ we have

$$\dim\left(\bigcup_{g_0 \in S_0} \varphi^{-1}(gg_0^{-1}) \times \psi^{-1}(g_0)\right) = \dim X + \dim Y - \dim\underline{G}. \tag{3.2}$$

By the geometric irreducibility of X we have $\dim\psi^{-1}(S_l) \leq \dim X - 1$ for every $l > 0$, and therefore by (3.1) and (3.2)

$$\dim(\varphi * \psi)^{-1}(g) \leq \dim X + \dim Y - \dim\underline{G} + k - 1. \qquad \square$$

3.2. Singularity properties of schemes and morphisms via their jets

In [**Mus01**] and [**EM04**], it was shown that various singularity properties of a scheme X can be detected in terms of its jet schemes:

THEOREM 3.10. *If X is a geometrically irreducible, local complete intersection K-variety then the following hold:*

1. *If X is normal, then $J_m(X)$ is equidimensional for every $m \geq 1$ (and hence $\dim J_m(X) = (m+1)\dim X$) if and only if X has log canonical singularities.*
2. *$J_m(X)$ is geometrically irreducible for all $m \geq 1$ if and only if X has rational singularities.*
3. *$J_m(X)$ is normal for all $m \geq 1$ if and only if X has terminal singularities.*

This motivates us to introduce the following definitions:

DEFINITION 3.11. Let $\varphi : X \to Y$ be a K-morphism of smooth varieties.

(1) φ is called *jet-flat* (resp. *jet-normal/jet-(FGI)*) if $J_m(\varphi) : J_m(X) \to J_m(Y)$ is flat (resp. normal/(FGI)) for every $m \geq 0$.
(2) φ is called *(FLCS)* (resp. *(FTS)*) if φ is flat, with normal fibers of log canonical singularities (resp. terminal singularities).

We now prove the following relative version of Theorem 3.10.

COROLLARY 3.12. *Let $\varphi : X \to Y$ be an (FGI)-morphism of smooth geometrically irreducible K-varieties X and Y.*

(1) *If φ is normal then φ is (FLCS) if and only if φ is jet-flat.*
(2) *φ is (FRS) if and only if φ is jet-(FGI).*
(3) *φ is (FTS) if and only if φ is jet-normal.*

REMARK 3.13. We have $(FTS) \Rightarrow (FRS) \Rightarrow (FLCS)$.
For the proof of Corollary 3.12 we need the following useful lemma.

LEMMA 3.14. *Let $\varphi : X \to Y$ be a dominant morphism between smooth geometrically irreducible K-varieties.*

(1) *Assume that φ is smooth and (FGI). Then φ is also jet-(FGI).*
(2) *Assume that φ has a geometrically irreducible generic fiber. Then for any $m \in \mathbb{N}$ the jet morphism $J_m(\varphi) : J_m(X) \to J_m(Y)$ has a geometrically irreducible generic fiber.*

PROOF. By [**Sta18**, Lemmas 37.25.5, 37.25.6] and by generic flatness, the generic fiber of $J_m(\varphi)$ is geometrically irreducible if and only if $J_m(\varphi)$ is (FGI) over an open set in $J_m(Y)$. By generic smoothness, it is enough to prove (1). Write $d = \dim X - \dim Y$. We may assume that φ factors through $X \overset{\psi}{\to} Y \times \mathbb{A}^d \overset{\mathrm{pr}}{\to} Y$, where ψ an étale map, and pr the projection. We can then write $J_m(\varphi) = J_m(\mathrm{pr}) \circ J_m(\psi)$, with $J_m(\psi) : J_m(X) \to J_m(Y \times \mathbb{A}^d)$ an étale map and $J_m(\mathrm{pr}) : J_m(Y \times \mathbb{A}^d) \to J_m(Y)$ is the projection.

Write $x \in J_m(\mathbb{A}^d) \simeq \mathbb{A}^{d(m+1)}$ as $x = (t, t')$ where $t \in \mathbb{A}^d$ and where $t' \in \mathbb{A}^{dm}$ corresponds to the various derivatives. Since ψ is étale, by Lemma 2.5, the following is a base change diagram:

$$\begin{array}{ccccc} J_m(X) & \overset{J_m(\psi)}{\longrightarrow} & J_m(Y) \times J_m(\mathbb{A}^d) & \overset{\Phi}{\longrightarrow} & J_m(Y) \times \mathbb{A}^{dm} \\ \pi_0^m \downarrow & & \pi_0^m \downarrow & & \pi \downarrow \\ X & \overset{\psi}{\longrightarrow} & Y \times \mathbb{A}^d & \overset{\mathrm{pr}}{\longrightarrow} & Y \end{array},$$

where Φ is defined by $\Phi(y, y', t, t') = (y, y', t')$ (for $(y, y') \in J_m(Y)$), and $\pi(y, y', t') = \pi_0^m(y, y') = y$. Notice that for any $(y, y', t') \in J_m(Y) \times \mathbb{A}^{dm}$, the scheme theoretic fiber $J_m(X)_{(y,y',t'),\Phi \circ J_m(\psi)}$ is a base change of the fiber $X_{y,\varphi}$ by a field extension, so it is geometrically irreducible.

For any $(y, y') \in J_m(Y)$ consider the map $f : J_m(X)_{(y,y'),J_m(\varphi)} \to \{(y, y')\} \times \mathbb{A}^{dm}$ by $(x, x') \mapsto \Phi \circ J_m(\psi)(x, x')$. Then it is a flat morphism with geometrically irreducible target, and such that all fibers are geometrically irreducible, as they can be identified with $J_m(X)_{(y,y',t'),\Phi \circ J_m(\psi)}$. Applying the following lemma to f, we deduce that $J_m(X)_{(y,y'),J_m(\varphi)}$ is geometrically irreducible for any $(y, y') \in J_m(Y)$, as required.

LEMMA 3.15 (see [**Sta18**, Lemma 5.8.12]). *Let $f : X_1 \to X_2$ be a continuous map between topological spaces, such that f is open, X_2 is irreducible, and there* *exists a dense collection of points $y \in X_2$ such that $f^{-1}(y)$ is irreducible. Then X_1 is irreducible.* □

PROOF OF COROLLARY 3.12. (1) and (3) follow from Corollary 2.17 and Theorem 3.10. One direction of (2) follows from Theorem 3.10. Assume that φ is (FRS). Then it follows from Theorem 3.10 that $J_m(\varphi) : J_m(X) \to J_m(Y)$ is (FGI) over $Y \subseteq J_m(Y)$. It is left to prove that it is (FGI).

By Corollary 2.17 $J_m(\varphi)$ is flat. By [**Mus01**, Proposition 1.5] $J_m(\varphi)$ is flat with fibers which are locally integral at each $x \in X \subseteq X_m$. By Corollary 2.17, all of the fibers of $J_m(\varphi)$ are locally integral, and in particular all geometrically irreducible components of the any fiber of $J_m(\varphi)$ are disjoint.

Since X, Y are smooth and $J_m(\varphi)$ is flat, we have $J_m(X^{\mathrm{sm},\varphi}) \simeq J_m(X)^{\mathrm{sm},J_m(\varphi)}$. Indeed, by writing the Jacobian of $J_m(\varphi)$, one can see that $J_m(\varphi)$ is a submersion at $(x, x') \in J_m(X)$ only if φ is a submersion at $x \in X$, and in the other direction, $\varphi|_{X^{\mathrm{sm},\varphi}}$ is smooth and thus $J_m(\varphi)|_{J_m(X^{\mathrm{sm},\varphi})}$ is smooth.

Let $(x, x') \in J_m(X)$ and set $(y, y') := J_m(\varphi)(x, x')$. Let Z be any geometrically irreducible component of $J_m(X)_{(y,y'),J_m(\varphi)}$. By Lemma 3.14, $J_m(\varphi)|_{X^{\mathrm{sm},\varphi}}$ has geometrically irreducible fibers, and thus $J_m(X)_{(y,y'),J_m(\varphi)} \cap J_m(X)^{\mathrm{sm},J_m(\varphi)}$ is geometrically irreducible. But since $J_m(\varphi)$ is flat with reduced fibers, it follows that $J_m(X)_{(y,y'),J_m(\varphi)} \cap J_m(X)^{\mathrm{sm},J_m(\varphi)}$ is dense in $J_m(X)_{(y,y'),J_m(\varphi)}$ which implies that $J_m(X)_{(y,y'),J_m(\varphi)}$ is geometrically irreducible. □

We would like to prove a version of Corollary 3.12 for non-(FGI) morphisms:

COROLLARY 3.16. *Let $\varphi : X \to Y$ be an (FRS) morphism between smooth geometrically irreducible K-varieties X and Y. Then for any $m \in \mathbb{N}$, $y \in \varphi(X)$ and $\tilde{y} \in J_m(Y)$ with $\pi_0^m(\tilde{y}) = y$, the schemes $J_m(X)_{\tilde{y},J_m(\varphi)}$ and $X_{y,\varphi}$ have the same number of geometrically irreducible components.*

PROOF. We may assume that Y is affine. Let us first assume that φ is smooth. For any $(y, y^{(1)}, \ldots, y^{(m)}) \in J_m(Y)$, the truncation map $\pi_{m-1}^m : J_m(X) \to J_{m-1}(X)$ induces a morphism

$$\psi_m : J_m(X)_{(y,y^{(1)},\ldots,y^{(m)}),J_m(\varphi)} \to J_{m-1}(X)_{(y,y^{(1)},\ldots,y^{(m-1)}),J_{m-1}(\varphi)}.$$

Note that the fibers of ψ_m are isomorphic to $\mathbb{A}^{\dim X - \dim Y}$. Indeed, we may assume that X is affine, and write $J_m(\varphi) = (\varphi, \varphi^{(1)}, \ldots, \varphi^{(m)})$. For any

$$(x, x^{(1)}, \ldots, x^{(m-1)}) \in J_{m-1}(X)_{(y,y^{(1)},\ldots,y^{(m-1)}),J_{m-1}(\varphi)},$$

we have

$$\psi_m^{-1}(x, x^{(1)}, \ldots, x^{(m-1)}) \simeq \{z : \varphi^{(m)}(x, x^{(1)}, \ldots, x^{(m-1)}, z) = y^{(m)}\}.$$

Since $\varphi^{(m)}$ is linear in $x^{(m)}$ where $x, x^{(1)}, \ldots, x^{(m-1)}$ are parameters, we are done. Restricting to each geometrically irreducible component of $X_{y,\varphi}$, and applying Lemma 3.15 to ψ_m, we deduce by induction on m that $J_m(X)_{(y,y^{(1)},\ldots,y^{(m)}),J_m(\varphi)}$ and $X_{y,\varphi}$ have the same number of geometrically irreducible components.

Since we proved the corollary for smooth morphisms, we deduce that

$$J_m(X)_{\tilde{y},J_m(\varphi)} \cap J_m(X^{\mathrm{sm},\varphi})$$

has the same number of geometrically irreducible components as $X_{y,\varphi} \cap X^{\mathrm{sm},\varphi}$, which is the same as in $X_{y,\varphi}$. But we have seen that $J_m(X^{\mathrm{sm},\varphi})$ intersects any irreducible component of $J_m(X)_{(y,y'),J_m(\varphi)}$ and they are all disjoint, so we are done. □

The m-th jet scheme $J_m(\underline{G})$ of an algebraic K-group $\underline{G}$ has a natural structure of an algebraic K-group, where the multiplication map $\mathrm{m}_{J_m(\underline{G})} : J_m(\underline{G}) \times J_m(\underline{G}) \to J_m(\underline{G})$ is just the m-th jet $J_m(\mathrm{m}_{\underline{G}})$ of the usual multiplication map $\mathrm{m}_{\underline{G}}$. In particular, we have the following lemma:

LEMMA 3.17. *Let X_1, X_2 be K-schemes, $\underline{G}$ an algebraic K-group, and let $\{\varphi_i : X_i \to \underline{G}\}_{i=1}^2$ be morphisms. Then $J_m(\varphi_1) * J_m(\varphi_1) = J_m(\varphi_1 * \varphi_2)$.*

COROLLARY 3.18. *Let X_1, X_2, X_3 be smooth geometrically irreducible K-varieties, and let $\{\varphi_i : X_i \to \underline{G}\}_{i=1}^3$ be dominant morphisms such that $\varphi_1 : X_1 \to \underline{G}$ is jet-flat. Then $\varphi_1 * \varphi_2$ is (FRS) and $\varphi_1 * \varphi_2 * \varphi_3$ is (FTS).*

PROOF. By Proposition 3.6 and Lemma 3.17 it follows that $J_m(\varphi_1 * \varphi_2)$ is flat with reduced fibers. In particular, $\left(J_m(X_1 \times X_2)_{(g,0)}\right)^{\mathrm{sm}} \simeq (J_m((X_1 \times X_2)_g))^{\mathrm{sm}}$ is dense in $J_m((X_1 \times X_2)_g)$ for any $g \in \underline{G}$. We have also seen that $J_m(X_1 \times X_2)^{\mathrm{sm}, J_m(\varphi_1 * \varphi_2)} \simeq J_m((X_1 \times X_2)^{\mathrm{sm}, \varphi_1 * \varphi_2})$ so

$$\begin{aligned}\left(J_m(X_1 \times X_2)_{(g,0)}\right)^{\mathrm{sm}} &= J_m(X_1 \times X_2)_{(g,0)} \cap J_m(X_1 \times X_2)^{\mathrm{sm}, J_m(\varphi_1 * \varphi_2)} \\ &= J_m((X_1 \times X_2)_g \cap (X_1 \times X_2)^{\mathrm{sm}, \varphi_1 * \varphi_2}) \\ &= J_m((X_1 \times X_2)_g^{\mathrm{sm}}).\end{aligned}$$

By [**Mus01**, Proposition 1.4], we deduce that $(X_1 \times X_2)_g$ has rational singularities. Proposition 3.6 and Corollary 3.12 imply that $\varphi_1 * \varphi_2 * \varphi_3$ is (FTS). □

3.3. Singularity properties of word maps: First observations

We have seen in Proposition 3.4 that a dominant morphism $\varphi : X \to \underline{G}$ becomes flat after $\dim \underline{G}$ self-convolutions, and one cannot hope for better bounds in this generality. Since word maps furnish an abundant of symmetries, one would expect their singularity properties improve much faster under self-convolution than those of a general morphism. For example, since word maps are conjugate invariant, the locus of any given singularity property is conjugate invariant as well. Indeed, in [**LST19**] it was shown that any word map $\varphi_w : \underline{G}^r \to \underline{G}$ becomes flat after $O(\ell(w)^4)$ self-convolutions, for every semisimple group $\underline{G}$, where $\ell(w)$ is the length of w. Similarly, one would expect improved bounds for Lie algebra word maps.

In Chapters 4 and 5 we provide bounds on the number of self-convolutions of Lie algebra word maps required to obtain flat and (FRS) morphisms. While the task of providing effective bounds for the (FRS) property of group word maps is not yet finished, we can still say something on the singularities of the fibers of word maps (see Chapter 7). For this, we would like to introduce quantitative ways to measure how bad are the singularities of a word map, or how far is a word map from being an (FRS) morphism. We further discuss the (FGI) property which plays an important role in the probabilistic applications of this paper. We first recall a result of [**LST19**]:

THEOREM 3.19 ([**LST19**, Lemma 2.4]). *Let $\underline{G}$ be a simply connected, semisimple K-algebraic group and let $w_1 \in F_{r_1}$ and $w_2 \in F_{r_2}$ be non-trivial words. Then $\varphi_{w_1} * \varphi_{w_2}$ has a geometrically irreducible generic fiber.*

From Theorem 3.19 we deduce the following corollary:

COROLLARY 3.20. *Let $\underline{G}$ be a simply connected, semisimple K-algebraic group. Let $\{1 \neq w_i \in F_{r_i}\}_{i=1}^3$ and assume that $\varphi_{w_1} * \varphi_{w_2}$ is flat. Then $\varphi_{w_1} * \varphi_{w_2} * \varphi_{w_3}$ is (FGI).*

PROOF. Set $r = \sum_{i=1}^{3} r_i$ and $w = w_1 * w_2 * w_3$. Given $g \in \underline{G}$, let $\pi_3 : (\underline{G}^r)_{g,\varphi_w} \to \underline{G}^{r_3}$ be the restriction to $(\underline{G}^r)_{g,\varphi_w}$ of the projection $\underline{G}^r \to \underline{G}^{r_3}$. Note that φ_w is flat (via Proposition 3.3), so $(\underline{G}^r)_{g,\varphi_w}$ is Cohen-Macaulay. Moreover, for any $g_3 \in \underline{G}^{r_3}$ we have

$$((\underline{G}^r)_{g,\varphi_w})_{g_3,\pi_3} \simeq \left(\underline{G}^{r_1+r_2}\right)_{g\cdot\varphi_{w_3}(g_3)^{-1},\varphi_{w_1}*\varphi_{w_2}},$$

so the fibers of π_3 are of dimension $\dim\underline{G}(r_1 + r_2 - 1) = \dim(\underline{G}^r)_{g,\varphi_w} - \dim\underline{G}^{r_3}$. By miracle flatness, π_3 is flat, and thus open. By Theorem 3.19, there is a dense collection of fibers of π_3 which are geometrically irreducible, so by Lemma 3.15 we deduce that $(\underline{G}^r)_{g,\varphi_w}$ is geometrically irreducible, as required. □

The following Corollary is an analogue of Theorem 3.19 for jet schemes, and is a direct application of Lemma 3.14 combined with Theorem 3.19.

COROLLARY 3.21. *Let $\underline{G}$ be a simply connected, semisimple K-algebraic group and let $w_1 \in F_{r_1}$ and $w_2 \in F_{r_2}$ be non-trivial words. Then for each $m \in \mathbb{N}$, the jet map $J_m(\varphi_{w_1*w_2}) : J_m(\underline{G})^{r_1+r_2} \to J_m(\underline{G})$ has a geometrically irreducible generic fiber.*

Finally we would like to introduce quantitative versions of flatness and jet-flatness, which we will use in Chapters 5 and 7.

DEFINITION 3.22. Let $\varphi : X \to Y$ be a morphism between geometrically irreducible smooth K-varieties.

(1) φ is called *ε-flat* if for every $x \in X$ we have $\dim X_{\varphi(x),\varphi} \leq \dim X - \varepsilon\dim Y$.
(2) φ is called *ε-jet flat* if $J_m(\varphi)$ is ε-flat for every $m \in \mathbb{N}$.

Note that a 1-flat (resp. 1-jet flat) morphism is just flat (resp. jet-flat).

LEMMA 3.23. *Let $\varphi : X \to Y$ be a morphism between geometrically irreducible smooth K-varieties. Then φ is ε-jet flat if and only if* $\operatorname{lct}(X, X_{\varphi(x),\varphi}) \geq \varepsilon\dim Y$ *for all $x \in X$.*

PROOF. From Theorem 2.6 we see that if φ is ε-jet flat then $\operatorname{lct}(X, X_{\varphi(x),\varphi}) \geq \varepsilon\dim Y$ for all $x \in X$, and that if the latter holds then $J_m(\varphi)$ is ε-flat at $X \subseteq J_m(X)$. By upper semi-continuity of the dimension of the fiber on the source (see [**Gro67**, Theorem 13.1.3]) and using the $\mathbb{G}_m$-action on $J_m(X)$, we get that $J_m(\varphi)$ is ε-flat. □

Motivated by Definition 3.22, we give the following definition for word maps:

DEFINITION 3.24. A word $w \in F_r$ is called *ε-flat* (resp. *ε-jet flat*) if for every simple algebraic K-group $\underline{G}$, the map $\varphi_w : \underline{G}^r \to \underline{G}$ is ε-flat (resp. ε-jet flat).

REMARK 3.25. A similar notion of ε-(jet) flatness can be defined for Lie algebra words $w \in \mathcal{L}_r$ (resp. algebra words $w \in \mathcal{A}_r$), by varying over all simple K-Lie algebras $\mathfrak{g}$ (resp. $\{M_n\}_{n\in\mathbb{N}}$) for which $\varphi_w : \mathfrak{g}^r \to \mathfrak{g}$ (resp. $\varphi_w : M_n^r \to M_n$) is non-trivial.

The following observation is useful for showing that a morphism is ε-flat (resp. ε-jet flat):

LEMMA 3.26. *Let X be a geometrically irreducible smooth K-variety, $\underline{G}$ be a connected algebraic K-group and let $\varphi : X \to \underline{G}$ be a morphism. Assume that φ^{*t} is flat (resp. jet-flat). Then φ is $\frac{1}{t}$-flat (resp. $\frac{1}{t}$-jet flat).*

PROOF. It is enough to prove the claim for flatness. Assume that φ is not $\frac{1}{t}$-flat. Then there exists $g \in \underline{G}$ such that $\dim X_{g,\varphi} > \dim X - \frac{1}{t}\dim \underline{G}$. In particular, $X^t_{g^t,\varphi^{*t}}$ contains $(X_{g,\varphi})^t$ so we have

$$\dim X^t_{g^t,\varphi^{*t}} > \dim X^t - \dim \underline{G},$$

and hence φ^{*t} is not flat. □

CHAPTER 4

Lie algebra word maps

4.1. Definitions and discussion of Lie algebra word maps

DEFINITION 4.1. Let $\mathcal{A}_r$ (resp. $\mathcal{L}_r$) be the free associative K-algebra (resp. the free K-Lie algebra) on a finite set $\{X_1, \ldots, X_r\}$.

(1) We refer to an element w of $\mathcal{A}_r$ (resp. of $\mathcal{L}_r$) as an *algebra word* (resp. a *Lie algebra word*).
(2) For a matrix algebra $M_n := \mathrm{Mat}_{n\times n}$ (resp. a Lie algebra $\mathfrak{g}$), the word w in $\mathcal{A}_r$ (resp. in $\mathcal{L}_r$) induces a *matrix word map* (resp. a *Lie word map*) $\varphi_w : M_n^r \to M_n$ (resp. $\varphi_w : \mathfrak{g}^r \to \mathfrak{g}$).
(3) Given two words w_1 and w_2 (either algebra words or Lie algebra words), we may define their (additive) convolution $w_1 * w_2 := w_1 + w_2$.
(4) Note that both $\mathcal{A}_r$ and $\mathcal{L}_r$ have natural gradations. We define the *degree* of a word w as the maximal grade $d \in \mathbb{N}$ in which the image of w is non-trivial. A word w is said to be *d-homogeneous* if it lies in the d-th grade.

REMARK 4.2. We can also define a multiplicative convolution operation. Unlike the addition map, the multiplication map $M_n \times M_n \to M_n$ is not smooth, so one does not expect the multiplicative convolution operation to improve singularity properties of morphisms.

In this section we would like to study the smoothing effect of the additive convolution operation on Lie (and matrix) algebra word maps, and show that under very mild assumptions, any Lie algebra word map $\varphi_w : \mathfrak{g}^r \to \mathfrak{g}$ where $\mathfrak{g}$ is a semisimple Lie algebra, becomes flat and eventually (FRS) after a number of convolutions $N(w)$ which is independent of $\mathfrak{g}$.

Having no assumptions on w turns out to be too optimistic. It is a classical theorem of Borel, that given a non-trivial word $w \in F_r$, the corresponding word map $\varphi_w : \underline{G}^r \to \underline{G}$ on any semisimple algebraic group $\underline{G}$ is dominant [**Bor83**, Theorem B]. Unlike the group case, a non-trivial Lie algebra word might induce a trivial word map on certain simple Lie algebras.

EXAMPLE 4.3 ([**BGKP12**, Example 3.8], [**Raz73**]). The word

$$w(X,Y,Z) = [[[[Z,Y],Y],X],Y] - [[[[Z,Y],X],Y],Y]$$

is non-trivial, but induces the zero map on $\mathfrak{sl}_2$.

Aside of this pathology, the situation is not too bad. In [**BGKP12**], a Lie algebra analogue of Borel's theorem was proven:

THEOREM 4.4 ([**BGKP12**, Theorem 3.2]). *Let $w \in \mathcal{L}_r$ be a Lie algebra word, and assume $\varphi_w : \mathfrak{sl}_2^r \to \mathfrak{sl}_2$ is non-trivial. Then $\varphi_w : \mathfrak{g}^r \to \mathfrak{g}$ is dominant for any semisimple K-Lie algebra $\mathfrak{g}$.*

In addition, one can observe the following:

PROPOSITION 4.5. *Let $\mathfrak{g}$ be a semisimple K-Lie algebra, and $w \in \mathcal{L}_r$ a Lie algebra word. Assume $\varphi_w : \mathfrak{g}^r \to \mathfrak{g}$ is non-zero. Then φ_w is generating.*

PROOF. It is enough to prove the statement when $\mathfrak{g}$ is simple and when $K = \mathbb{C}$. Let $W := \mathrm{span}_{\mathbb{C}}\varphi_w(\mathfrak{g}(\mathbb{C})^r)$ and let $\underline{G}$ be a simply connected group with $\mathrm{Lie}(\underline{G}) = \mathfrak{g}$. Notice that for any $g \in \underline{G}(\mathbb{C})$ and $X_1, \ldots, X_r \in \mathfrak{g}(\mathbb{C})$,

$$\varphi_w(\mathrm{Ad}_g X_1, \ldots, \mathrm{Ad}_g X_r) = \mathrm{Ad}_g \varphi_w(X_1, \ldots, X_r),$$

so W is $\underline{G}(\mathbb{C})$-invariant. By the simplicity of $\mathfrak{g}(\mathbb{C})$ we must have $W = \mathfrak{g}(\mathbb{C})$. □

CONJECTURE 4.6 ([**BGKP12**, Question 3.1]). *Let $\mathfrak{g}$ be a semisimple K-Lie algebra, and $w \in \mathcal{L}_r$. Assume $\varphi_w : \mathfrak{g}^r \to \mathfrak{g}$ is non-zero (i.e. $\varphi_w(\mathfrak{g}(\overline{K})^r) \neq \{0\}$). Then φ_w is dominant.*

By Theorem 3.5 and Lemma 3.8 (or by [**GH19**, Theorem 1.7]), any generating map into a vector space becomes dominant, and eventually (FRS), after sufficiently many self-convolutions. Thus, the assumption that $\varphi_w : \mathfrak{g}^r \to \mathfrak{g}$ is non-zero is sufficient for our purposes. Now, from the Amitsur-Levitzki theorem, we can deduce that any Lie algebra word map is non-trivial if the rank of $\mathfrak{g}$ is large enough:

THEOREM 4.7 (Amitsur-Levitzki, [**AL50**]). *Let R be a commutative ring. Any polynomial identity on $M_d(R)$ is of degree at least $2d$.*

COROLLARY 4.8.

(1) *Any $0 \neq w \in \mathcal{L}_r$ such that $\varphi_w : \mathfrak{sl}_d^r \to \mathfrak{sl}_d$ is trivial is of degree at least $2d$.*
(2) *Any $0 \neq w \in \mathcal{L}_r$ such that $\varphi_w : \mathfrak{g}^r \to \mathfrak{g}$ is trivial on a simple classical Lie algebra $\mathfrak{g}$ of rank d is of degree at least $2d$.*

PROOF. (1) follows easily from Amitsur-Levitzki and (2) follows from (1) since each of $\mathfrak{so}_{2d}, \mathfrak{so}_{2d+1}$ and $\mathfrak{sp}_{2d}$ contains $\mathfrak{sl}_d$ as a Lie subalgebra. □

4.2. Formulation of the main theorems

We turn to formulate some of our main results.

THEOREM 4.9. *Let $\{w_i \in \mathcal{L}_{r_i}\}_{i\in\mathbb{N}}$ be a collection of Lie algebra words of degree at most $d \in \mathbb{N}$. Then there exists $0 < C < 10^6$, such that for any semisimple K-Lie algebra $\mathfrak{g}$ for which $\{\varphi_{w_i}\}_{i\in\mathbb{N}}$ are non-trivial when restricted to each simple constituent of $\mathfrak{g}$, we have the following:*

(1) *If $m \geq Cd^3$ then $\varphi_{w_1} * \cdots * \varphi_{w_m}$ is flat.*
(2) *If $m \geq Cd^3 + 2$ then $\varphi_{w_1} * \cdots * \varphi_{w_m}$ is flat with normal fibers.*
(3) *If $m \geq Cd^4$ then $\varphi_{w_1} * \cdots * \varphi_{w_m}$ is (FRS).*
(4) *If $m \geq Cd^4 + 2$ then $\varphi_{w_1} * \cdots * \varphi_{w_m}$ is (FTS).*

Note that by Corollary 4.8, if the rank of every simple constituent of a semisimple Lie algebra $\mathfrak{g}$ is larger than $d/2$, then every Lie algebra word of degree at most d induces a non-trivial Lie algebra word map on $\mathfrak{g}$.

An analogue of Theorem 4.9 holds also for matrix word maps:

THEOREM 4.10. *Let $\{w_i \in \mathcal{A}_{r_i}\}_{i\in\mathbb{N}}$ be a collection of matrix words of degree at most $d \in \mathbb{N}$. Then for any $n \in \mathbb{N}$ such that $\{\varphi_{w_i} : M_n^r \to M_n\}_{i\in\mathbb{N}}$ are generating, the assertions of Theorem 4.9 hold.*

Theorems 4.9 and 4.10 combined with Lemma 3.26, imply that (Lie) algebra word maps are ε-jet flat, for $\varepsilon \sim d^{-4}$. In fact, we can achieve slightly better lower bounds on the ε.

THEOREM 4.11. *Let $w_1 \in \mathcal{A}_r$ and $w_2 \in \mathcal{L}_r$ be words of degree at most d. Then w_1 is $\frac{1}{25d^3}$-jet flat and w_2 is $\frac{1}{2 \cdot 10^5 d^3}$-jet flat.*

The arguments appearing in the proofs of Theorems 4.9 and 4.10 generalize the arguments used in [**AA16**, Theorem 2.1], in which a version of Theorem 4.9 was proved for the specific case of the commutator Lie algebra word.

The main approach for proving Theorem 4.9 is as follows; we first reduce to the case of self-convolutions of d-homogeneous Lie algebra word maps (of pure type). Given a d-homogeneous word w and Lie algebra $\mathfrak{g}$, we apply a series of $\mathbb{G}_m$-degenerations to the word map $\varphi_w : \mathfrak{g}^r \to \mathfrak{g}$, such that the resulting map ψ_w is much easier to analyze. We then use Corollary 2.12 to conclude φ_w^{*t} is flat (resp. (FRS)), given we managed to show ψ_w^{*t} is flat (resp. (FRS)), for some $t \in \mathbb{N}$. The price of degenerating a map is that its regularity properties might diminish. Our main goal is to apply a series of carefully chosen degenerations to φ_w, such that the number of convolutions of ψ_w needed to get a flat (resp. (FRS)) morphism is uniformly bounded over all semisimple Lie algebras $\mathfrak{g}$, and behaves nicely with respect to d.

To achieve this goal, we construct a certain 'combinatorial gadget' which encodes the word map $\varphi_w : \mathfrak{g}^r \to \mathfrak{g}$, and generalizes the construction given in [**AA16**] in the case of the commutator map. We then translate the flatness and (FRS) properties, as well as the convolution operation and the degeneration methods we use to properties of and operations on this gadget.

Let us first explain the combinatorial construction in the case of the commutator map (see also [**AA16**]).

4.3. Combinatorial picture in the case of the commutator word

As before, let $[\cdot,\cdot]$ be the commutator Lie algebra word, and by abusing notation, given a Lie algebra $\mathfrak{g}$ we denote by $[\cdot,\cdot] : \mathfrak{g} \times \mathfrak{g} \to \mathfrak{g}$ the corresponding Lie algebra word map. For simplicity, we work over the field of complex numbers. The proof that there exists a number t such that $[\cdot,\cdot]^{*t} : \mathfrak{g}^{2t} \to \mathfrak{g}$ is (FRS) for every semisimple Lie algebra $\mathfrak{g}$ can be explained in six main steps:

(1) **Construct a polygraph from a classical Lie algebra $\mathfrak{g}$;** We attach to each simple Lie algebra $\mathfrak{g}$ a polygraph $\Gamma_{\mathfrak{g}} = (\mathcal{I}, \mathcal{J}, \mathcal{S})$, which is a triplet consisting of finite sets $\mathcal{I}, \mathcal{J}$ and a subset $\mathcal{S} \subseteq \mathcal{I}^{\{2\}} \times \mathcal{J}$, where $\mathcal{I}^{\{2\}}$ denotes the set of subsets of $\mathcal{I}$ of size 2. The difference between a polygraph and a graph is that a polygraph has edges of different *types*, where each type is an element of $\mathcal{J}$. Here, the vertices $\{e_i\}_{i\in\mathcal{I}}$ of the polygraph correspond to a Chevalley basis of $\mathfrak{g}$ and the types $\mathcal{J}$ correspond to elements of the dual basis $\{\widehat{e}_l\}_{l\in\mathcal{J}}$ of $\mathfrak{g}^*$. The edges $\mathcal{S}$ are pairs $(\{e_i, e_j\}, \widehat{e}_l)$, such that $\widehat{e}_l([e_i, e_j]) \neq 0$.

For example, taking $\mathfrak{g} = \mathfrak{sl}_d$, one can choose the basis $\mathcal{B} = \{E_{i,j}\}_{i\neq j} \cup \{H_{i,i}\}$, where $(E_{i,j})_{i',j'} = \delta_{i,i'}\delta_{j,j'}$ (here $\delta_{x,y}$ stands for the Kronecker delta function), and $H_{i,i} = [E_{i,i+1}, E_{i+1,i}]$. The set of types $\mathcal{J}$ corresponds to the basis dual to $\mathcal{B}$, denoted $\widehat{\mathcal{B}} = \{\hat{E}_{i,j}\} \cup \{\hat{H}_{i,i}\}$. The vertices $E_{i,j}$ and

$E_{j,k}$ satisfy $\hat{E}_{i,k}([E_{i,j}, E_{j,k}]) = \hat{E}_{i,k}(E_{i,k}) \neq 0$, so $(\{E_{i,j}, E_{j,k}\}, \hat{E}_{i,k}) \in \mathcal{S}$ represents an edge of type $\hat{E}_{i,k}$.

(2) **Describe $[\cdot,\cdot]^{*t} : \mathfrak{g}^{2t} \to \mathfrak{g}$ in terms of the constructed polygraph $\Gamma_{\mathfrak{g}}$;** Let $\Gamma_{\mathfrak{g}} = (\mathcal{I}, \mathcal{J}, \mathcal{S})$ as above. We have the following identification of vector spaces $\mathfrak{g}^{2t} \simeq \mathfrak{g} \otimes V \simeq V^{\mathcal{I}}$, where V is a $2t$-dimensional vector space. Let $(a_{i,j,l} := \hat{e}_l([e_i, e_j]))_{i,j\in\mathcal{I}, l\in\mathcal{J}}$ be the structure coefficients of $\mathfrak{g}$ with respect to $\mathcal{I}$ and $\mathcal{J}$. Then under the identifications above, $[\cdot,\cdot]^{*t}$ corresponds to the map $\Phi^{*t}_{\mathrm{comm}} : V^{\mathcal{I}} \to \mathbb{C}^{\mathcal{J}}$ which is given by

$$\Phi^{*t}_{\mathrm{comm}}((v_i)_{i\in\mathcal{I}}) = \left(\sum_{i,j\in\mathcal{I}} a_{i,j,l} \cdot \left(\sum_{k=1}^{t} v_{i,2k-1} v_{j,2k}\right)\right)_{l\in\mathcal{J}}$$

$$= \left(\sum_{i<j\in\mathcal{I}} a_{i,j,l} \cdot \left(\sum_{k=1}^{t} v_{i,2k-1} v_{j,2k} - v_{j,2k-1} v_{i,2k}\right)\right)_{l\in\mathcal{J}},$$

where each $v_i \in V$ is of the form $(v_{i,1}, \dots, v_{i,2t})$. Note that the sum $\sum_{k=1}^{t} v_{i,2k-1} v_{j,2k} - v_{j,2k-1} v_{i,2k}$ is just a standard symplectic form on a $2t$-dimensional symplectic space $(V, \langle \cdot, \cdot \rangle)$. The map $\Phi^{*t}_{\mathrm{comm}} : V^{\mathcal{I}} \to \mathbb{C}^{\mathcal{J}}$ can then be written as

$$\Phi^{*t}_{\mathrm{comm}}((v_i)_{i\in\mathcal{I}}) = \left(\sum_{i<j\in\mathcal{I}} a_{i,j,l} \cdot \langle v_i, v_j \rangle\right)_{l\in\mathcal{J}}.$$

This operation can be considered more abstractly.

Definition 4.12 ([**AA16**, Definition 2.22]). Given a polygraph $\Gamma = (\mathcal{I}, \mathcal{J}, \mathcal{S})$, an ordering $<$ of $\mathcal{I}$, a function $a : \mathcal{S} \to \mathbb{C}^{\times}$ and a symplectic space $(V, \langle \cdot, \cdot \rangle)$, we define a map $\Phi_{\Gamma,<,a,V} : V^{\mathcal{I}} \to \mathbb{C}^{\mathcal{J}}$ by

$$\Phi_{\Gamma,<,a,V}((v_i)_{i\in\mathcal{I}}) = \left(\sum_{i<j\in\mathcal{I}} a(\{i,j\}, l) \cdot \langle v_i, v_j \rangle\right)_{l\in\mathcal{J}}.$$

Let Γ be a polygraph and V be a symplectic space. We say that the pair (Γ, V) is *(FRS)* if for all $<$ and a as above, the map $\Phi_{\Gamma,<,a,V}$ is (FRS).

Notice that the (FRS) property of $[\cdot,\cdot]^{*t} : \mathfrak{g}^{2t} \to \mathfrak{g}$ is equivalent to the (FRS) property of the polygraph $\Gamma_{\mathfrak{g}} = (\mathcal{I}, \mathcal{J}, \mathcal{S})$. It is also important to note that taking convolution powers of $[\cdot,\cdot]$ (or Φ_{comm}), translates to choosing a larger symplectic space, where t self-convolutions, correspond to assigning a $2t$-dimensional symplectic space to each element of $\mathcal{I}$.

(3) **Relate the elimination method to certain combinatorial operations on the polygraph;** Let $\Phi_{\Gamma,<,a,V}$ be the map induced by a polygraph Γ. Let $\omega \in \mathbb{Z}^{\mathcal{I}}$ be a vector, which we call an *$\mathcal{I}$-weight*. An $\mathcal{I}$-weight ω induces an $\mathcal{I}^{\{2\}}$-weight $\widetilde{\omega} : \mathcal{I}^{\{2\}} \to \mathbb{Z}$ by $\widetilde{\omega}(\{i_1, i_2\}) = \omega(i_1) + \omega(i_2)$. We furthermore define $\mathrm{gr}_\omega(\Gamma) := (\mathcal{I}, \mathcal{J}, \mathrm{gr}_\omega \mathcal{S})$ by

$$\mathrm{gr}_\omega \mathcal{S} = \{(\{i_1, i_2\}, l) \in \mathcal{S} : \forall (\{j_1, j_2\}, l) \in \mathcal{S} \text{ we have } \widetilde{\omega}(i_1, i_2\}) \leq \widetilde{\omega}(j_1, j_2\})\}.$$

An application of the elimination method (Corollary 2.15) gives the following.

COROLLARY 4.13 ([**AA16**, Corollary 2.28], elimination for polygraphs). *Let* $\Gamma := (\mathcal{I}, \mathcal{J}, \mathcal{S})$ *be a polygraph and* V *be a symplectic space. Let* $\omega \in \mathbb{Z}^{\mathcal{I}}$ *be an* $\mathcal{I}$*-weight. If* $(\mathrm{gr}_\omega(\Gamma), V)$ *is (FRS), then so is* (Γ, V).

Notice that this elimination procedure introduces a new polygraph $\mathrm{gr}_\omega(\Gamma)$ which has the same vertices but eliminates some of the edges, leaving only the edges $\mathrm{gr}_\omega \mathcal{S}$. Furthermore, this operation does not increase $\dim V$, as no convolutions are involved.

(4) **Reduction from a polygraph to a graph;** It turns out that for the special case of polygraphs induced from a classical Lie algebra $\mathfrak{g}$, one can use the elimination method twice, and obtain a new polygraph, Γ_2, which has a unique edge of each type $\widehat{e_l} \in \mathcal{J}$ (see [**AA16**, Section 2.6.2]). In this situation, the polygraph Γ_2 is just a graph. Since graphs $\Gamma = (\mathcal{V}, \mathcal{E})$, where $\mathcal{V}$ denotes the set of vertices, and $\mathcal{E}$ the edges, are special cases of polygraphs, we can define a map $\Phi_{\Gamma,V} : V^{\mathcal{V}} \to \mathbb{C}^{\mathcal{E}}$ as before. The graph $\Gamma = (\mathcal{V}, \mathcal{E})$ is then (FRS) if $\Phi_{\Gamma,V}$ is.

(5) **A method of coloring and reduction to an edge;** Let $\Gamma = (\mathcal{V}, \mathcal{E})$ be a graph, let M be a finite set, which we think of as our set of *colors*, and let $\omega : \mathcal{V} \to \mathbb{Z}^M$ be a choice of an M-tuple of weights to each vertex. We call such ω a *coloring function*. As in the case of polygraphs, ω induces a weight function on $\mathcal{V}^{\{2\}}$ by $\widetilde{\omega}(\{v_1, v_2\}) := \omega(v_1) + \omega(v_2)$. We say that ω is *admissible* if for each edge $\varepsilon = \{v_1, v_2\} \in \mathcal{E}$, the tuple $\widetilde{\omega}(\varepsilon) \in \mathbb{Z}^M$ has a unique minimum. We say that an edge ε has color $i \in M$ if for each $j \in M$ we have $\widetilde{\omega}(\varepsilon)_i \leq \widetilde{\omega}(\varepsilon)_j$. Notice that if ω is admissible, then each edge is colored in a unique color. This means we can define a new graph $\mathrm{gr}_{\omega,i}\Gamma := (\mathcal{V}, \mathrm{gr}_{\omega,i}\mathcal{E})$ where $\mathrm{gr}_{\omega,i}\mathcal{E} = \{\varepsilon \in \mathcal{E} : \varepsilon \text{ has color } i\}$. The following corollary relates the (FRS) property of Γ to the (FRS) property of each of the $\mathrm{gr}_{\omega,i}\Gamma$:

COROLLARY 4.14 ([**AA16**, Corollary 2.33], coloring method). *Let* $\Gamma := (\mathcal{V}, \mathcal{E})$ *be a graph,* M *a finite set and* $\omega : \mathcal{V} \to \mathbb{Z}^M$ *an admissible coloring. Let* $\{V_i\}_{i \in M}$ *be symplectic spaces and* $V = \bigoplus_{i \in M} V_i$. *Suppose that* $(\mathrm{gr}_{\omega,i}\Gamma, V_i)$ *is (FRS). Then so is* (Γ, V).

It is important to note that unlike the elimination method for polygraphs (Corollary 4.13), the coloring method comes with a price. Namely, we may divide our graph $\Gamma := (\mathcal{V}, \mathcal{E})$ to simpler graphs $\mathrm{gr}_{\omega,i}\Gamma$ with fewer edges, and prove each is (FRS) separately, but we "pay" by enlarging our symplectic space, which means more convolutions of $[\cdot, \cdot] : \mathfrak{g}^2 \to \mathfrak{g}$. Thus, the coloring method can be seen as applying M self-convolutions, followed by an elimination of edges.

In [**AA16**, Section 2.6.3], Avni and Aizenbud showed that using 3 colors one can reduce the graph Γ_2 from Step 4 to a disjoint union of trees (i.e. a forest), each with degree bounded by a number D. The bound on the degree D depends on the type of $\mathfrak{g}$ (e.g. for $\mathfrak{sl}_n$, taking $D = 3$ suffices).

After reducing to a forest of maximal degree D, one can easily reduce to a disjoint union of edges with the use of $2(D-1)$ colors (see [**AA16**, Theorem 2.35]).

(6) **Proof for the case of an edge;** Given a pair (Γ, V),where Γ is a graph consisting of a single edge, the map $\Phi_{\Gamma,V}$ has a very simple description. It

corresponds to the map $\Psi^{*\frac{\dim V}{2}}: \mathbb{A}^{2\dim V} \to \mathbb{A}^1$ where $\Psi(x_1, x_2, x_3, x_4) = x_1x_2 - x_3x_4$. Since Ψ can be easily shown to be (FRS), it is enough to take $\dim V = 2$.

Taking $\mathfrak{sl}_n$ for example, the above steps are as follows. We first construct the polygraph $\Gamma = (\mathcal{I}, \mathcal{J}, \mathcal{S})$ corresponding to $[\cdot,\cdot] : \mathfrak{sl}_n^2 \to \mathfrak{sl}_n$. We then reduce the polygraph Γ to a graph Γ_2. Three colors are needed to reduce Γ_2 to a forest Γ_3 of degree≤ 3. Then, using four more colors, one can reduce to an edge, which is (FRS) if V is at least two dimensional. Altogether, we deduce that if $\dim V = 2 \cdot 4 \cdot 3 = 24$ then (Γ, V) is (FRS) and hence also $[\cdot,\cdot]^{*12} : \mathfrak{sl}_n^{24} \to \mathfrak{sl}_n$ is (FRS), regardless of n.

4.3.1. Generalizing to word maps: Motivation. We would like to prove Theorems 4.9 and 4.10 using similar tools as above. It would be convenient to reduce our problem to word maps of homogeneous degree d, and to the special case of self-convolutions of word maps (i.e. to consider $\varphi_{w_1}^{*m}$ rather then $\varphi_{w_1} * \cdots * \varphi_{w_m}$).

A different combinatorial gadget is needed in order to deal with words w of homogeneous degree $d > 2$; the edges in the polygraph in the case of the commutator word, correspond to commutator relations between various pairs of elements in a chosen basis $\mathcal{B}$ of the Lie algebra $\mathfrak{g}$. In the case of a d-homogeneous word, the edges should encode relations between d elements in $\mathcal{B}$, so one should consider hypergraphs instead of graphs, where edges now consist of sets (or multisets) of d vertices.

As discussed in the introduction, the main challenge is to degenerate the hypergraph such that the resulting map will be simpler to analyze, with similar singularity properties (or only slightly worse), and generating. This is much more involved in the general case than the commutator case. Certain pathologies such as in Example 4.3 appear when the rank of the Lie algebra is comparable to the degree d of w. Such situations will require a more careful treatment.

Before we construct the combinatorial gadget, we first make some reductions.

4.4. Reduction to self-convolutions of homogeneous word maps of pure type

DEFINITION 4.15. An element in $\mathcal{L}_r(X_1, \ldots, X_r)$ of the form

$$[[\ldots[[X_{s_1}, X_{s_2}], X_{s_3}], \ldots], X_{s_d}]$$

is called a *(left-normed) Lie monomial of degree d,* and it is denoted $X_S := [X_{s_1}, \ldots, X_{s_d}]$, for $S = (s_1, \ldots, s_d) \in [r]^d$. Note that here, X_s counts as a Lie monomial of degree 1.

The Jacobi identity and the anti-commutativity imply the following:

FACT 4.16 (e.g. [**Chi06**, Lemma 3.3]). *Any d-homogeneous Lie algebra word $w \in \mathcal{L}_r$ can be written as a linear combination of left-normed Lie monomials of degree d.*

DEFINITION 4.17. A word $w \in \mathcal{L}_r(X_1, \ldots, X_r)$ is called *d-homogeneous of pure type*, if it is of the form

$$w(X_1, \ldots, X_r) = \sum_{S \in \mathrm{MS}^{-1}(S_0)} c_S X_S,$$

where $c_S \in K$, S_0 is an element of $[r]^{(d)}$, the set of multisets of $[r]$ of size d, and $\mathrm{MS} : [r]^d \to [r]^{(d)}$ is the forgetful map from d-tuples to d-multisets. In this case we say that w is *of pure type* S_0.

PROPOSITION 4.18. *It is enough to prove Theorem 4.9 for d-homogeneous word maps of pure type.*

The proof of Proposition 4.18 relies on the following lemma:

LEMMA 4.19. *Let* $\{\varphi_i = (\varphi_{i,1}, \ldots, \varphi_{i,m}) : \mathbb{A}^{n_i} \to \mathbb{A}^m\}_{i=1}^N$ *be polynomial maps, and let* $\widetilde{\varphi}_{i,j} \in K[x_1, \ldots, x_{n_i}]$ *be the homogeneous part of* $\varphi_{i,j}$ *of maximal degree. Assume that* $\widetilde{\varphi}_1 * \ldots * \widetilde{\varphi}_N : \mathbb{A}^{n_1} \times \ldots \times \mathbb{A}^{n_N} \to \mathbb{A}^m$ *is flat (resp. (FRS)) at* $(0, \ldots, 0)$. *Then* $\varphi_1 * \ldots * \varphi_N$ *is flat (resp. (FRS)).*

PROOF. We prove the lemma in the case $N = 1$, the general case is similar. Write

$$\varphi_{1,i}(x_1, \ldots, x_n) = \sum_{k=0}^{l_i} P_{i,k}(x_1, \ldots, x_n),$$

where $P_{i,k}$ is a homogeneous polynomial of degree k, and P_{i,l_i} is non-zero. Now, define a map $\psi = (t, \psi_1, \ldots, \psi_m) : \mathbb{A}^1 \times \mathbb{A}^n \to \mathbb{A}^1 \times \mathbb{A}^m$ by

$$\psi_i(t, x_1, \ldots, x_n) = \sum_{k=0}^{l_i} t^{l_i - k} P_{i,k}(x_1, \ldots, x_n).$$

Then ψ is an $\mathbb{A}^1$-family of morphisms parameterized by t, where $\psi(1) = \varphi$ and $\psi(0) = \widetilde{\varphi}$. Consider the following action of $\mathbb{G}_m$; for $s \in K^\times$ set $s.(t, x_1, \ldots, x_n) = (st, sx_1, \ldots, sx_n)$ and $s.(t, y_1, \ldots, y_m)) = (st, s^{l_1} y_1, \ldots, s^{l_m} y_m)$. Note that ψ is $\mathbb{G}_m$-equivariant, as for any $s \in K^\times$,

$$\begin{aligned}
\psi_i(st, sx_1, \ldots, sx_n) &= \sum_{k=0}^{l_i} s^{l_i - k} t^{l_i - k} P_{i,k}(sx_1, \ldots, sx_n) \\
&= s^{l_i} \sum_{k=0}^{l_i} t^{l_i - k} P_{i,k}(x_1, \ldots, x_n) \\
&= s.\psi_i(t, x_1, \ldots, x_n).
\end{aligned}$$

Now, take $(a_1, \ldots, a_n) \in K^n$ and choose the section $f : \mathbb{A}^1 \to \mathbb{A}^1 \times \mathbb{A}^n$ defined by $f(t) = (t, ta_1, \ldots, ta_n)$. Clearly f is $\mathbb{G}_m$-equivariant, so by Corollary 2.12 φ is flat (resp. (FRS)) at $(a_1, \ldots, a_n)$. □

REMARK 4.20. Note that if $\varphi : \mathbb{A}^n \to \mathbb{A}^m$ is a homogeneous map which is flat (resp. (FRS)) at $(0, \ldots, 0)$, then it is flat (resp. (FRS)).

PROOF OF PROPOSITION 4.18. Let $\{w_i \in \mathcal{L}_{r_i}\}_{i \in \mathbb{N}}$ be a collection of Lie algebra words of degree at most d, and set $w = w_{i_1} * \cdots * w_{i_m}$. For each i, let $\widetilde{w}_i$ be the homogeneous part of maximal degree of w_i. By the above lemma, in order to prove that $\varphi_w = \varphi_{w_{i_1}} * \cdots * \varphi_{w_{i_m}}$ is flat (resp. (FRS)) it is enough to show that $\varphi_{\widetilde{w}_{i_1}} * \cdots * \varphi_{\widetilde{w}_{i_m}}$ is flat (resp. (FRS)). We have therefore reduced Theorem 4.9 to the case of homogeneous words.

To reduce to the case of homogeneous words of pure type, we use the elimination method (Corollary 2.15). Given a homogeneous word $v \in \mathcal{L}(X_1, \ldots, X_r)$ of degree d, we choose a weight function $\omega(X_i) = (d+1)^i$, that is, we assign the same weight

$(d+1)^i$ to all of the variables in the i-th copy of $\mathfrak{g}$ which is represented by X_i. This degenerates v to a homogeneous word of pure type, since two Lie monomials X_I and X_J of degree d satisfy $\deg_\omega(X_I) = \deg_\omega(X_J)$ if and only if $\mathrm{MS}(I) = \mathrm{MS}(J)$. $\square$

We would now like to further reduce to the case of self-convolutions:

PROPOSITION 4.21. *It is enough to prove Theorem 4.9 in the case of self-convolutions of homogeneous word maps of pure type.*

The proof of Proposition 4.21 follows directly from the following proposition:

PROPOSITION 4.22. *Let $\{\varphi_i : X_i \to V\}_{i\in I}$ be a collection of morphisms, with X_i smooth and geometrically irreducible for each i, and V a K-vector space. Assume that there exists $t \in \mathbb{N}$ such that φ_i^{*t} is flat (resp. (FRS)) for every i. Then $\varphi_1 * \cdots * \varphi_{2t}$ is flat (resp. (FRS)).*

PROOF. Write $\psi := \varphi_1 * \cdots * \varphi_{2t}$ and

$$X := \prod_{i=1}^{2t} X_i.$$

The case of (FRS) was proved in [**GH19**, Lemma 7.3]. For flatness, we prove the claim in the case $K = \mathbb{Q}$, and the reduction to $\mathbb{Q}$ is similar to [**GH19**, Section 6].

By choosing a $\mathbb{Z}$-model, we may assume that ψ is defined over $\mathbb{Z}$. Denote by $\mu_{X(\mathbb{F}_p)}$ and $\mu_{V(\mathbb{F}_p)}$ the uniform probability measures on $X(\mathbb{F}_p)$ and $V(\mathbb{F}_p)$, and let $\tau_{\psi,p} := \psi_*(\mu_{X(\mathbb{F}_p)})$, and $\tau_{i,p} := \varphi_{i*}(\mu_{X_i(\mathbb{F}_p)})$. There exists p_0 such that for every $p > p_0$, the map φ_i^{*t} is flat over $\mathbb{F}_p$ for every i. By Corollary 8.4, there exists $C > 0$ such that for every $p > p_0$ and every i, we have $\left\|\tau_{i,p}^{*t}\right\|_\infty < C$. Therefore also

$$\|\mathcal{F}(\tau_{i,p})\|_{2t}^t = \left\|\mathcal{F}(\tau_{i,p})^t\right\|_2 = \left\|\tau_{i,p}^{*t}\right\|_2 \le \left\|\tau_{i,p}^{*t}\right\|_\infty < C.$$

By a generalization of Hölder's inequality:

$$\|\tau_{\psi,p}\|_\infty \le \|\mathcal{F}(\tau_{\psi,p})\|_1 = \left\|\prod_{i=1}^{2t}\mathcal{F}(\tau_{i,p})\right\|_1 \le \prod_{i=1}^{2t}\|\mathcal{F}(\tau_{i,p})\|_{2t} < C^2.$$

By Corollary 8.4, this implies that ψ is flat so we are done. $\square$

REMARK 4.23. Proposition 4.22 is also true if we replace V with a connected algebraic K-group. Instead of the commutative Fourier transform, one has to work with the non-commutative one. For a very similar argument, see [**GH21**, Proposition 5.3].

4.5. Combinatorial description of homogeneous word maps

We would like to generalize the construction discussed in Section 4.3 to the case of homogeneous Lie algebra word maps of any degree.

4.5.1. The combinatorial gadget.

DEFINITION 4.24.

(1) A *d-polyhypergraph (or d-PH)* is a triplet $\Gamma := (\mathcal{I}, \mathcal{J}, \mathcal{S})$ consisting of finite sets $\mathcal{I}, \mathcal{J}$ and a subset $\mathcal{S} \subseteq \mathcal{I}^{(d)} \times \mathcal{J}$, where $\mathcal{I}^{(d)}$ is the set of multisets of $\mathcal{I}$ of size d. An element $i \in \mathcal{I}$ is called a *vertex* of Γ. An element $\{i_1, \dots, i_d\} \in \mathcal{I}^{(d)}$ is called a *hyperedge of* Γ, if there exists $l \in \mathcal{J}$ such that $(\{i_1, \dots, i_d\}, l) \in \mathcal{S}$. A hyperedge $\{i_1, \dots, i_d\} \in \mathcal{I}^{(d)}$ is *of type* $l \in \mathcal{J}$

if $(\{i_1, \dots, i_d\}, l) \in \mathcal{S}$. In particular, $\mathcal{S}$ is the collection of all tuples consisting of a hyperedge together with a type.

(2) Given a d-polyhypergraph $\Gamma = (\mathcal{I}, \mathcal{J}, \mathcal{S})$, an assignment $\mathrm{V} = (V(i))_{i\in\mathcal{I}}$ of a vector space $V(i)$ to each vertex $i \in \mathcal{I}$, and a collection $\{\tau_\gamma\}_{\gamma\in\mathcal{S}}$ of d-homogeneous multilinear forms

$$\tau_{\{i_1,\dots,i_d\},l} : \bigoplus_{i\in\{i_1,\dots,i_d\}} V(i) \to K,$$

we define a map $\Phi_{\Gamma,\mathrm{V},\{\tau_\gamma\}} : \bigoplus_{i\in\mathcal{I}} V(i) \to K^{\mathcal{J}}$ by

$$\Phi_{\Gamma,\mathrm{V},\{\tau_\gamma\}}(v) = \left(\sum_{(\{i_1,\dots,i_d\},l)\in\mathcal{S}} \tau_{\{i_1,\dots,i_d\},l}(v) \right)_{l\in\mathcal{J}}. \tag{4.1}$$

(3) Let Γ be a d-PH. A triplet $(\Gamma, \mathrm{V}, \{\tau_\gamma\})$ is called an *assignment* of Γ. If all of the $V(i)$ are of the same dimension N, we write $(\Gamma, V, \{\tau_\gamma\})$, where V is just an N-dimensional vector space.

(4) We say that the assignment $(\Gamma, \mathrm{V}, \{\tau_\gamma\})$ is flat (resp. (FRS)) if the map $\Phi_{\Gamma,\mathrm{V},\{\tau_\gamma\}}$ is flat (resp. (FRS)).

(5) Let $\Gamma = (\mathcal{I}, \mathcal{J}, \mathcal{S})$ be a d-PH. Assume that

$$\mathcal{I} = \bigsqcup_{u=1}^{N} \mathcal{I}_u, \mathcal{J} = \bigsqcup_{u=1}^{N} \mathcal{J}_u$$

and that

$$\mathcal{S} = \bigsqcup_{u=1}^{N} \mathcal{S}_u,$$

where

$$\mathcal{S}_u := \mathcal{S} \cap \left(\mathcal{I}_u^{(d)} \times \mathcal{J}_u\right).$$

In this case we say that Γ is a *disjoint union of* $\{\Gamma_u\}_{u\in[N]}$, where $\Gamma_u = (\mathcal{I}_u, \mathcal{J}_u, \mathcal{S}_u)$.

REMARK 4.25. We further make the assumption that

$$\tau_{\{i_1,\dots,i_d\},l}((\lambda_i v_i)_{i\in\mathcal{I}}) = \left(\prod_{i=1}^{d} \lambda_i^{k_i}\right) \tau_{\{i_1,\dots,i_d\},l}((v_i)_{i\in\mathcal{I}}),$$

where $k_i := |\{j \in [d] : i_j = i\}|$.

Let us give a concrete example for a d-PH induced from a d-homogeneous word map $\varphi_w : \mathfrak{g}^r \to \mathfrak{g}$, which is the main object of study in this section and in Chapter 5.

4.5.2. Attaching a polyhypergraph to a homogeneous Lie algebra word map. Assume that K is algebraically closed. Let $\mathfrak{g}$ be a simple Lie algebra, $\mathfrak{t}$ a Cartan subalgebra, and write $\mathfrak{g} = \mathfrak{t} \oplus \bigoplus_{\alpha\in\Sigma(\mathfrak{g},\mathfrak{t})} \mathfrak{g}_\alpha$, where $\Sigma(\mathfrak{g}, \mathfrak{t})$ is the corresponding root system. Let $\triangle \subseteq \Sigma(\mathfrak{g}, \mathfrak{t})$ be the set of simple roots. We choose a Chevalley basis $\mathcal{B} = \{h_\alpha\}_{\alpha\in\triangle} \cup \{e_\beta\}_{\beta\in\Sigma(\mathfrak{g},\mathfrak{t})}$, such that $e_\beta \in \mathfrak{g}_\beta$, $h_\alpha = [e_\alpha, e_{-\alpha}] \in \mathfrak{t}$, and $[e_\beta, e_{-\beta}] \in \sum_{\gamma\in\triangle} \mathbb{Z}h_\gamma$, for any $\alpha \in \triangle$ and $\beta \in \Sigma(\mathfrak{g}, \mathfrak{t})$. Now let $w \in \mathcal{L}_r(X_1, \dots, X_r)$

be a d-homogeneous word and let $\varphi_w : \mathfrak{g}^r \to \mathfrak{g}$ be its corresponding Lie algebra word map. Recall from Fact 4.16 that w can be written as

$$w(X_1, \ldots, X_r) = \sum_{S \in [r]^d} c_S X_S,$$

for constants c_S. Set $\mathcal{I} = \mathcal{B}$, and let $\mathcal{J} = \widehat{\mathcal{B}}$ be the dual basis of $\mathcal{B}$. Consider $\varphi_w^{*t} : \mathfrak{g}^{rt} \to \mathfrak{g}$, let

$$\{X_{s,k} = \sum_{i \in \mathcal{I}} a_{i,s,k} e_i\}_{(s,k) \in [r] \times [t]}$$

be tr elements in $\mathfrak{g}$, and write $\mathrm{MS} : \mathcal{I}^d \to \mathcal{I}^{(d)}$ for the forgetful map from d-tuples to d-multisets. Then

$$\begin{aligned}
\varphi_w^{*t}(X_{1,1}, \ldots, X_{r,t}) &= \sum_{k=1}^{t} \varphi_w(X_{1,k}, \ldots, X_{r,k}) \\
&= \sum_{k=1}^{t} \sum_{S \in [r]^d} c_S [X_{s_1,k}, \ldots, X_{s_r,k}] \\
&= \sum_{k=1}^{t} \sum_{S \in [r]^d} c_S [\sum_{i \in \mathcal{I}} a_{i,s_1,k} e_i, \ldots, \sum_{i \in \mathcal{I}} a_{i,s_d,k} e_i] \\
&= \sum_{k=1}^{t} \sum_{I \in \mathcal{I}^{(d)}} \sum_{(i_1,\ldots,i_d) \in \mathrm{MS}^{-1}(I)} \sum_{S \in [r]^d} c_S \prod_{u=1}^{d} a_{i_u,s_u,k} [e_{i_1}, \ldots, e_{i_d}].
\end{aligned}$$

Rewriting this equation gives the following lemma.

LEMMA 4.26. *Let V be an rt-dimensional vector space. Under the identification $\mathfrak{g}^{rt} \simeq \mathfrak{g} \otimes V \simeq V^{\mathcal{I}}$, the map $\varphi_w^{*t} : \mathfrak{g}^{rt} \simeq V^{\mathcal{I}} \to K^{\mathcal{J}} \simeq \mathfrak{g}$ is given by*

$$\varphi_w^{*t}((v_i)_{i \in \mathcal{I}}) = \left(\sum_{k=1}^{t} \sum_{\{i_1,\ldots,i_d\} \in \mathcal{I}^{(d)}} \tau_{\{i_1,\ldots,i_d\},l}((v_i)_{i \in \mathcal{I}}) \right)_{l \in \mathcal{J}}$$

where $v_i = (v_{i,s,k})_{s \in [r], k \in [t]}$ and

$$\tau_{\{i_1,\ldots,i_d\},l}((v_i)_{i \in \mathcal{I}}) = \sum_{(i'_1,\ldots,i'_d) \in \mathrm{MS}^{-1}(\{i_1,\ldots,i_d\})} \sum_{S \in [r]^d} c_S \prod_{u=1}^{d} v_{i'_u,s_u,k} \widehat{e}_l([e_{i'_1}, \ldots, e_{i'_d}]).$$

REMARK 4.27. Given an embedding $\mathfrak{g} \hookrightarrow \mathfrak{gl}_N$, we may replace $[X,Y]$ with $X \cdot Y - Y \cdot X$ and write any Lie algebra word map $\varphi_w : \mathfrak{g}^r \to \mathfrak{g}$ as a restriction to $\mathfrak{g}$ of a matrix word map $\varphi_{\widetilde{w}} : M_N^r \to M_N$, with

$$\widetilde{w} = \sum_{S \in [r]^d} d_S X_{s_1} \cdot \cdots \cdot X_{s_d}$$

for some constants d_S.

EXAMPLE 4.28. Let $\mathfrak{g}$ be a simple Lie algebra, let $w(X_1, X_2) = [[X_1, X_2], X_2]$, let $t = 1$, and consider $\varphi_w^{*t} = \varphi_w$. In particular $\mathfrak{g}^{rt} = \mathfrak{g}^2 \simeq \mathfrak{g} \otimes V \simeq V^{\mathcal{I}}$ for V a two dimensional vector space. We have the following:

$$
\begin{aligned}
(\varphi_w((v_i)_{i\in\mathcal{I}}))_l &= \widehat{e}_l\left(\varphi_w\left(\sum_{i\in\mathcal{I}} v_{i,1}e_i, \sum_{i\in\mathcal{I}} v_{i,2}e_i\right)\right)\\
&= \widehat{e}_l\left([\sum_{i\in\mathcal{I}} v_{i,1}e_i, \sum_{i\in\mathcal{I}} v_{i,2}e_i, \sum_{i\in\mathcal{I}} v_{i,2}e_i)]\right)\\
&= \sum_{\{i'_1,i'_2,i'_3\}\in\mathcal{I}^{(3)}} \sum_{(i_1,i_2,i_3)\in \mathrm{MS}^{-1}(\{i'_1,i'_2,i'_3\})} v_{i_1,1}v_{i_1,2}v_{i_3,2}\left(\hat{e}_l([e_{i_1},e_{i_2},e_{i_3}])\right).
\end{aligned}
$$

For example, taking $\{i'_1,i'_2,i'_3\} = \{1,2,2\}$, we have

$$\mathrm{MS}^{-1}(\{i'_1,i'_2,i'_3\}) = \{(1,2,2),(2,1,2),(2,2,1)\},$$

and

$$
\begin{aligned}
&\tau_{\{1,2,2\},l}(v_{1,1},v_{1,2},v_{2,1},v_{2,2})\\
&= v_{1,1}v_{2,2}^2\hat{e}_l([e_1,e_2,e_2]) + v_{2,1}v_{1,2}v_{2,2}\hat{e}_l([e_2,e_1,e_2]) + v_{2,1}v_{2,2}v_{1,2}\hat{e}_l([e_2,e_2,e_1])\\
&= \left(v_{1,1}v_{2,2}^2 - v_{2,1}v_{1,2}v_{2,2}\right)\hat{e}_l([e_1,e_2,e_2]).
\end{aligned}
$$

DEFINITION 4.29. Let $w \in \mathcal{L}_r$ be a d-homogeneous word, and $\mathfrak{g}$ a simple Lie algebra. Define the d-PH corresponding to $\mathfrak{g}$ and w denoted $\Gamma_{\mathfrak{g},w} = (\mathcal{I},\mathcal{J},\mathcal{S})$ as follows. Take $\mathcal{I}$ and $\mathcal{J}$, to be a chosen Chevalley basis $\mathcal{B} = \{e_i\}_{i\in\mathcal{I}}$, and its dual $\widehat{\mathcal{B}} = \{\widehat{e}_j\}_{j\in\mathcal{J}}$. Define $\mathcal{S} \subseteq \mathcal{I}^{(d)} \times \mathcal{J}$ by

$$\mathcal{S} := \{(\{i_1,\dots,i_d\},l) \in \mathcal{I}^{(d)} \times \mathcal{J} : \tau_{\{i_1,\dots,i_d\},l} \neq 0\},$$

with $\tau_{\{i_1,\dots,i_d\},l}$ as in Lemma 4.26. Notice that for each $l \in \mathcal{J}$ there exists at least one $\{i_1,\dots,i_d\}$ such that $\tau_{\{i_1,\dots,i_d\},l}$ is non-zero, as otherwise the map φ_w is not generating, contrary to our assumption. Let V be an tr-dimensional vector space. Then $(\Gamma_{\mathfrak{g},w}, V, \{\tau_\gamma\}_{\gamma\in\mathcal{S}})$ is an assignment of $\Gamma_{\mathfrak{g},w}$. Notice that the collection $\{\tau_\gamma\}_{\gamma\in\mathcal{S}}$ satisfies the demand of Remark 4.25.

4.5.3. Coloring and elimination for polyhypergraphs. We next describe two combinatorial procedures on a given assignment of a d-PH: the elimination method and the coloring method. The elimination method is a direct generalization of Corollary 4.13 (c.f. [**AA16**, Corollary 2.28]) to the case of polyhypergraphs. The coloring method was described in [**AA16**, Corollary 2.33] in the case of graphs, but can be easily generalized to polygraphs and polyhypergraphs.

Let $\Gamma = (\mathcal{I},\mathcal{J},\mathcal{S})$ be a d-PH. Let $\omega \in \mathbb{Z}^{\mathcal{I}}$ be a vector (termed an $\mathcal{I}$-weight). Then ω induces an $\mathcal{I}^{(d)}$-weight $\widetilde{\omega} : \mathcal{I}^{(d)} \to \mathbb{Z}$ by

$$\widetilde{\omega}(\{i_1,\dots,i_d\}) = \sum_{u=1}^{d} \omega(i_u).$$

We define $\mathrm{gr}_\omega(\Gamma) := (\mathcal{I},\mathcal{J},\mathrm{gr}_\omega\mathcal{S})$ by

$$\mathrm{gr}_\omega\mathcal{S} = \{(I,l) \in \mathcal{S} : \forall (I',l) \in \mathcal{S} \text{ we have } \widetilde{\omega}(I) \leq \widetilde{\omega}(I')\},$$

where $I, I' \in \mathcal{I}^{(d)}$ and $l \in \mathcal{J}$.

COROLLARY 4.30 (cf. [**AA16**, Corollary 2.28]). *Let* $(\Gamma = (\mathcal{I},\mathcal{J},\mathcal{S}), \mathrm{V}, \{\tau_\gamma\}_{\gamma\in\mathcal{S}})$ *be an assignment of a d-PH, and let* $\omega \in \mathbb{Z}^{\mathcal{I}}$ *be an* $\mathcal{I}$*-weight. If* $(\mathrm{gr}_\omega(\Gamma), \mathrm{V}, \{\tau_\gamma\})$ *is flat (resp. (FRS)), then so is* $(\Gamma, \mathrm{V}, \{\tau_\gamma\})$.

PROOF. By Corollary 2.15, it is enough to show that the symbol $\sigma_\omega(\Phi_{\Gamma,\mathrm{V},\{\tau_\gamma\}})$ is equal to $\Phi_{\mathrm{gr}_\omega\Gamma,\mathrm{V},\{\tau_\gamma\}}$. By our assumption on the forms $\{\tau_\gamma\}_{\gamma\in\mathcal{S}}$ (Remark 4.25) and by Equation (4.1) it is easy to see that the symbol $\sigma_\omega(\Phi_{\Gamma,\mathrm{V},\{\tau_\gamma\}})$ is indeed $\Phi_{\mathrm{gr}_\omega\Gamma,\mathrm{V},\{\tau_\gamma\}}$. □

REMARK 4.31 (Warning!). Consider the map $\varphi : \mathbb{A}^2 \to \mathbb{A}^2$ defined by $\varphi(x,y) = (y^2+x,x)$, and let $\omega = (1,1)$ be the weight which induces the standard degree of monomials. Note that φ is dominant, while $\sigma_\omega(\varphi)(x,y) = (x,x)$ is not even generating, so $\sigma_\omega(\varphi)$ will never become (FRS) after any number of convolutions. This phenomenon might occur when applying the elimination method, so we must choose our weight functions carefully.

Let $\Gamma := (\mathcal{I},\mathcal{J},\mathcal{S})$ be a d-PH. We describe the coloring method for Γ. Let M be a finite set, and let $\omega : \mathcal{I} \to \mathbb{Z}^M$ be a map which we call a *coloring function*. Define $\widetilde{\omega} : \mathcal{I}^{(d)} \to \mathbb{Z}^M$ by

$$\widetilde{\omega}(\{i_1,\dots,i_d\}) := \sum_{u=1}^{d} \omega(i_u),$$

and for any $m \in M$ set $\mathrm{gr}_{\omega,m}\Gamma := (\mathcal{I}, \mathrm{gr}_{\omega,m}\mathcal{J}, \mathrm{gr}_{\omega,m}\mathcal{S})$, where

$$\mathrm{gr}_{\omega,m}\mathcal{S} = \{(I,l) \in \mathcal{S} : \forall m' \in M \text{ we have } \widetilde{\omega}(I)_m \le \widetilde{\omega}(I)_{m'}\} \subseteq \mathcal{S}, \tag{4.2}$$

$$\mathrm{gr}_{\omega,m}\mathcal{J} = \{l \in \mathcal{J} : \exists I \in \mathcal{I}^{(d)} \text{ such that } (I,l) \in \mathrm{gr}_{\omega,m}\mathcal{S}\}.$$

We say that the coloring function ω is *admissible* if for any $l \in \mathcal{J}$, there exists $m \in M$ such that for any hyperedge $(I,l) \in \mathcal{S}$, the number $\widetilde{\omega}(I)_m$ is the unique minimum of the tuple $\widetilde{\omega}(I) \in \mathbb{Z}^M$. We have the following analogue of Corollary 4.14.

COROLLARY 4.32 (cf. [**AA16**, Corollary 2.33], the coloring method). *Let $\Gamma := (\mathcal{I},\mathcal{J},\mathcal{S})$ be a d-PH, let M be a finite set, let $\omega : \mathcal{I} \to \mathbb{Z}^M$ be an admissible coloring function and let $(\mathrm{gr}_{\omega,m}\Gamma, \mathrm{V}_m, \{\tau_\gamma^m\}_\gamma)$ be assignments of $\mathrm{gr}_{\omega,m}\Gamma$ for each $m \in M$. Suppose that $(\mathrm{gr}_{\omega,m}\Gamma, \mathrm{V}_m, \{\tau_\gamma^m\}_\gamma)$ is flat (resp. (FRS)) for each $m \in M$. Then so is $(\Gamma, \mathrm{V}, \{\tau_\gamma\}_\gamma)$, with $\mathrm{V}(i) := \bigoplus_{m=1}^{M} \mathrm{V}_m(i)$ and $\tau_\gamma = \bigoplus_{i=1}^{M} \tau_\gamma^m$ for each $\gamma \in \mathcal{S}$.*

PROOF. Set $\Gamma_M := (\mathcal{I}\times M, \mathcal{J}, \mathcal{S}_M)$, where $\mathcal{S}_M = \{((I,m),l) : (I,l) \in \mathcal{S}, m \in M\}$, to be the d-PH consisting of M copies of $\Gamma = (\mathcal{I},\mathcal{J},\mathcal{S})$. Then we define a weight vector $\omega' \in \mathbb{Z}^{\mathcal{I}\times M}$, by $\omega'(i,m) := \omega(i)_m$. Consider the assignment $(\Gamma_M, \widetilde{\mathrm{V}}, \{\widetilde{\tau}_{(\gamma,m)}\})$ of Γ_M, with $\widetilde{\mathrm{V}}(i,m) = \mathrm{V}_m(i)$ and $\widetilde{\tau}_{(\gamma,m)} := \tau_\gamma^m$ for any $m \in M$ and $\gamma \in \mathcal{S}$. Notice that $(\Gamma_M, \widetilde{\mathrm{V}}, \{\widetilde{\tau}_{(\gamma,m)}\})$ is flat (resp. (FRS)) if and only if $(\Gamma, \mathrm{V}, \{\tau_\gamma\})$ is flat (resp. (FRS)), as both polygraphs induce the same morphism. Notice that by the admissibility of ω, we get that $\mathrm{gr}_{\omega'}(\Gamma_M)$ is a disjoint union (in the sense of Definition 4.24) of the polyhypergraphs $\mathrm{gr}_{\omega,m}\Gamma$ for $1 \le m \le M$. In particular, the fiber

$$\Phi^{-1}_{\mathrm{gr}_{\omega'}(\Gamma_M),\widetilde{\mathrm{V}},\{\widetilde{\tau}_{(\gamma,m)}\}}(0)$$

is isomorphic to the product

$$\prod_{m\in M} \Phi^{-1}_{\mathrm{gr}_{\omega,m}\Gamma,\mathrm{V}_m,\{\tau_\gamma^m\}}(0).$$

The claim now follows from Corollary 4.30. □

REMARK 4.33. The coloring method is just a different manifestation of the elimination method, after first applying convolutions; let M be a finite set, $(t_m)_{m\in M} \in \mathbb{Z}_{>0}^{|M|}$, with

$$t = \sum_{m=1}^{|M|} t_m,$$

let $\varphi : \mathbb{A}^{n_2} \to \mathbb{A}^{n_1}$ be a morphism, and let $\omega : [n_2] \to \mathbb{Z}^M$ be a (coloring) function. Then the coloring method can be seen as first considering $\varphi^{*t} : \mathbb{A}^{n_2 t} \to \mathbb{A}^{n_1}$, and then applying the elimination method, by assigning the weight $\nu_\omega : [n_2] \times [t] \to \mathbb{Z}$ by $\nu_\omega(i,l) = \omega(i)_{m(l)}$, where $m(l) = m'$ if

$$\sum_{j=1}^{m'-1} t_j < l \leq \sum_{j=1}^{m'} t_j.$$

The admissibility of the coloring ω guarantees that $\sigma_{\nu_\omega}(\varphi^{*t})$ is a disjoint union of morphisms $\sigma_{\omega_m}(\varphi^{*t_m})$. In particular, if $\sigma_{\omega_m}(\varphi^{*t_m})$ is flat or (FRS) for every $m \in M$, then so is φ^{*t}.

DEFINITION 4.34. Let $\Gamma := (\mathcal{I}, \mathcal{J}, \mathcal{S})$ be a d-PH and let $(\Gamma, \mathrm{V}, \{\tau_\gamma\})$ be an assignment. Then $(\Gamma, \mathrm{V}^{\oplus t}, \{\tau_\gamma^{\oplus t}\})$ is called the *t-th self-convolution* of $(\Gamma, \mathrm{V}, \{\tau_\gamma\})$. In particular, $(\Gamma, \mathrm{V}, \{\tau_\gamma\})$ is said to be *flat* (resp. *(FRS)*) *after* t *convolutions* if $(\Gamma, \mathrm{V}^{\oplus t}, \{\tau_\gamma^{\oplus t}\})$ is flat (resp. (FRS)).

The following is a special case of a d-PH which will be particularly important to us, and is easier to analyze:

DEFINITION 4.35. A *d-hypergraph* Γ is a pair $\Gamma = (\mathcal{V}, \mathcal{E})$, where $\mathcal{V}$ a set of vertices and $\mathcal{E} \subseteq \mathcal{V}^{(d)}$ is the set of d-hyperedges, where $\mathcal{V}^{(d)}$ is the set of multisubsets of $\mathcal{V}$ of size d. Any d-hypergraph induces a natural d-PH

$$\mathcal{P}(\Gamma) := (\mathcal{V}, \mathcal{E}, \triangle\mathcal{E})$$

where $\triangle\mathcal{E} \subseteq \mathcal{V}^{(d)} \times \mathcal{E}$ embedded diagonally. An *assignment* of Γ is just an assignment of $\mathcal{P}(\Gamma)$. Similarly as for polyhypergraphs, we say that an assignment $(\Gamma, \mathrm{V}, \{\tau_I\}_{I\in\mathcal{E}})$ of Γ is (FRS) if $(\mathcal{P}(\Gamma), V, \{\tau_I\}_{I\in\mathcal{E}})$ is (FRS).

Summarizing the above two methods, the elimination method helps us eliminate edges of a given d-PH, and results in a d-PH with fewer edges. The coloring method consists of first constructing a larger d-PH Γ_M which corresponds to taking M self-convolutions and consists of M copies of our d-PH Γ, and then using the elimination method on Γ_M. The use of the coloring method is essential in following two cases:

(1) Once we manage to reduce a polyhypergraph to a hypergraph, the elimination method will usually not be effective.
(2) In case we are given a d-PH, denoted Γ, but we cannot guarantee that the elimination method will result in new polygraph $\mathrm{gr}_\omega(\Gamma)$ whose corresponding map $\Phi_{\mathrm{gr}_\omega \Gamma, \mathrm{V}, \{\tau_\gamma\}}$ is generating.

CHAPTER 5

Lie algebra word maps: Proof of the main theorems

In this section we prove Theorems 4.9, 4.10 and 4.11. By Proposition 4.21 we may consider self-convolutions of a single d-homogeneous word $w \in \mathcal{L}_r$ of pure type. Our main methodology is to encode $\varphi_w^{*t} : \mathfrak{g}^r \to \mathfrak{g}$ in terms of an assignment $(\Gamma_{\mathfrak{g},w}, V, \{\tau_\gamma\}_{\gamma\in\mathcal{S}})$ of a d-PH $\Gamma_{\mathfrak{g},w}$, as in Definition 4.29, with V a tr-dimensional vector space. Using the elimination and coloring methods, we reduce $\Gamma_{\mathfrak{g},w}$ to a simpler object, at the price of enlarging $\dim V$. Here, a simpler object is either a d-PH consisting of (disjoint unions of) a single hyperedge, or a d-PH with a small number of types (relative to d). A complication arises when $\mathrm{rk}(\mathfrak{g})$ is small. This requires special attention. The structure of this section is as follows:

- In Section 5.1 we discuss the main methods we use - the elimination and coloring methods. These methods are used repeatedly throughout Chapter 5.
- In Section 5.2 we deal with low rank simple Lie algebras (in particular with all exceptional Lie algebras) and with polyhypergraphs with a few types of hyperedges.
- Sections 5.3 and 5.5 deal with the case of $\mathfrak{g}=\mathfrak{sl}_n$ and the general case, respectively.
- In Section 5.4 we discuss matrix word maps, and prove Theorem 4.10.

5.1. Discussion of the main methods

Throughout Chapter 5 we use the elimination and coloring methods abundantly, for polyhypergraphs, hypergraphs and general polynomial maps. Our applications of the coloring and elimination methods can be clustered into three types:

(1) The ***averaging method***, using ω_{av}.
(2) The ***monomialization method***, using ω_{mon}.
(3) The ***level separation method***, using ω_{LS}.

Let us explain these procedures using a simple example, which is very similar to the analysis of the jets of the commutator map considered in Chapter 6. Consider the following map (which is the $2t$-th jet map of $(x, y) \mapsto xy$):

$$\varphi(x_0, y_0, x_1, y_1, \ldots, x_{2t}, y_{2t}) = (x_0y_0, x_0y_1 + x_1y_0, \sum_{j=0}^{2} x_jy_{2-j}, \ldots, \sum_{j=0}^{2t} x_jy_{2t-j}).$$

We first apply the weight $\omega_{\mathrm{av}}(x_u) = \omega_{\mathrm{av}}(y_u) = 3^u$. Notice that $\deg_{\omega_{\mathrm{av}}}(x_jy_{u-j}) = 3^j + 3^{u-j}$ so that $\deg_{\omega_{\mathrm{av}}}$ is minimized when j is as close as possible to $\frac{u}{2}$. More precisely,

$$\sigma_{\omega_{\mathrm{av}}}(\varphi)(x_0, y_0, x_1, y_1, \ldots, x_{2t}, y_{2t}) = (x_0y_0, x_0y_1 + x_1y_0, x_1y_1, \ldots, x_{t-1}y_t + x_ty_{t-1}, x_ty_t).$$

We see that ω_{av} not only eliminated most of the monomials, but also that for any u_1, u_2 such that $|u_1 - u_2| > 1$, the polynomials $\sigma_{\omega_{\mathrm{av}}}(\varphi)_{u_1}$ and $\sigma_{\omega_{\mathrm{av}}}(\varphi)_{u_2}$ have disjoint variables, while for example, in φ_{2t} all of the variables $x_0, y_0, \ldots, x_t, y_t$ are involved. We call the use of ω_{av} and its variations the ***averaging method***.

Let us now apply a different type of weight. We first order the variables by $\mathrm{ord}(x_u) = 2u + 1$, and $\mathrm{ord}(y_u) = 2u + 2$ such that $x_0 < y_0 < \ldots < x_{2t} < y_{2t}$. Set

$$\omega_{\mathrm{mon}}(x_u) := 3^{\mathrm{ord}(x_u)} \text{ and } \omega_{\mathrm{mon}}(y_u) := 3^{\mathrm{ord}(y_u)}.$$

Notice that $\deg_{\omega_{\mathrm{mon}}}(x_{u-1}y_u) > \deg_{\omega_{\mathrm{mon}}}(x_u y_{u-1})$ and therefore:

$$\psi := \sigma_{\omega_{\mathrm{mon}}}(\sigma_{\omega_{\mathrm{av}}}(\varphi)) = (x_0y_0, x_1y_0, x_1y_1, \ldots, x_ty_{t-1}, x_ty_t)$$

is a monomial map. The use of ω_{mon} is designed to degenerate a polynomial map to a monomial map, so we refer to the use of ω_{mon} and its variations as the ***monomialization method.*** In the case of polyhypergraphs, we will use ω_{mon} to turn a polyhypergraph into a hypergraph.

Since the map is already monomial, we cannot use the elimination method (at least without a change of variables). Instead, we apply the coloring method, by applying one convolution, and then use the elimination method. The fact that the various coordinates of ψ have small interactions with their neighbors (a feature we earned by using ω_{av}), makes this method effective. Explicitly, we consider

$$\begin{aligned}&\psi^{*2}(x_0, y_0, z_0, w_0, \ldots, x_{2t}, y_{2t}, z_{2t}, w_{2t})\\ &\quad= \psi(x_0, y_0, \ldots, x_{2t}, y_{2t}) + \psi(z_0, w_0, \ldots, z_{2t}, w_{2t})\end{aligned}$$

and then use the weights

$$\omega_{\mathrm{LS}}(x_u) = 2 \cdot 10^u, \ \omega_{\mathrm{LS}}(y_u) = 3 \cdot 10^u, \ \omega_{\mathrm{LS}}(z_u) = 3 \cdot 10^u, \quad \text{and } \omega_{\mathrm{LS}}(w_u) = 10^u.$$

Notice that $\deg_{\omega_{\mathrm{LS}}}(x_uy_u) > \deg_{\omega_{\mathrm{LS}}}(z_uw_u)$ but $\deg_{\omega_{\mathrm{LS}}}(x_{u+1}y_u) < \deg_{\omega_{\mathrm{LS}}}(z_{u+1}w_u)$ for any u. Therefore we get

$$\phi := \sigma_{\omega_{\mathrm{LS}}}(\psi * \psi) = (z_0w_0, x_1y_0, z_1w_1, \ldots, x_ty_{t-1}, z_tw_t).$$

Notice that each pair of coordinates of ϕ has disjoint variables, and hence $\phi^{-1}(0)$ is isomorphic to the product

$$\prod_{u=0}^{t} (z_uw_u)^{-1}(0) \times (x_{u+1}y_u)^{-1}(0).$$

We have thus reduced the problem to a single monomial x_uy_u. Note that $x_1x_2 + x_3x_4 = 0$ has rational singularities, so $\phi^{*2} = \sigma_{\omega_{\mathrm{LS}}}(\psi^{*4})$ is (FRS), which implies by Corollary 2.15 that φ^{*4} is (FRS). The last method we used is a special case of the coloring method (as it involves an application of convolution followed by elimination) which we call ***level separation***, as we use ω_{LS} to separate a given map into 'levels' which have pairwise disjoint variables.

5.1.1. Degenerations of jets of Lie algebra word maps. We next describe a certain manifestation of the degeneration techniques presented above, applied to the jets of Lie algebra word maps. Since a formal derivation of a Lie algebra word map is a Lie algebra word map (Lemma 5.1), and since the (FRS) property of a morphism is closely related to the flatness of its m-th jets (Corollary 3.12), this allows us to reduce certain questions on the (FRS) property of Lie algebra word maps to question about flatness of word maps or maps built out of bounded "layers" of word maps. For example, in Chapter 6 we use this method to reduce the question about the (FRS) property of the commutator map to a question about flatness of the commutator map. This is further used in Section 5.4.

Let $w \in \mathcal{L}_r$ be a word of degree $d \in \mathbb{N}$, and $\mathfrak{g}$ be a simple K-Lie algebra. Following the notation from Section 2.1, for any $m \in \mathbb{N}$, we consider the m-th jet maps $J_m(\varphi_w) : J_m(\mathfrak{g})^r \to J_m(\mathfrak{g})$ where $J_m(\mathfrak{g})$ is identified with $\mathfrak{g}^{m+1}$ as vector spaces. We can then write

$$J_m(\varphi_w)^{*t}(X_1, \ldots, X_{r,t}, \ldots, X_1^{(m)}, \ldots, X_{r,t}^{(m)}) = \left(\varphi_w^{*t}, (\varphi_w^{*t})^{(1)}, \ldots, (\varphi_w^{*t})^{(m)}\right),$$

where $(\varphi_w^{*t})^{(u)}$ is the u-th formal derivative of φ_w^{*t}.

LEMMA 5.1. *For each $t, u \in \mathbb{N}$, the map $(\varphi_w^{*t})^{(u)}$ is a d-homogeneous word map.*

PROOF. Embed $\mathfrak{g} \hookrightarrow \mathfrak{gl}_n$ and then by functoriality of J_m, it is enough to prove the claim for $\mathfrak{g} = \mathfrak{gl}_n$. Note that

$$(\varphi_w^{*t})^{(u)} = \left(\varphi_w^{(u)}\right)^{*t}$$

so we may assume $t = 1$. By Fact 4.16 we may assume that $w = X_S$ is a left-normed Lie monomial.

The formal derivative of any monomial of matrices

$$X_{s_1} \cdot X_{s_2} \cdot \ldots \cdot X_{s_k} \text{ is } \sum_{j=1}^{k} X_{s_1} \cdot \ldots \cdot X_{s_j}^{(1)} \cdot \ldots \cdot X_{s_k}.$$

In particular, by induction on d it can be shown that

$$X_S^{(1)} = \sum_{k=1}^{d} [X_{s_1}, \ldots, X_{s_{k-1}}, X_{s_k}^{(1)}, X_{s_{k+1}}, X_{s_d}]. \tag{5.1}$$

By induction on u, we deduce that $\varphi_w^{(u)}$ is a d-homogeneous word map. □

DEFINITION 5.2. Recall that a word is of pure type $S_0 \in [r]^{(d)}$, if it can be written as

$$\sum_{S \in \mathrm{MS}^{-1}(S_0)} c_S X_S,$$

and note that any word w can be written as a sum of words of pure type

$$w = \sum_{S_0 \in [r]^{(d)}} w_{S_0} = \sum_{S_0 \in [r]^{(d)}} \left(\sum_{S \in \mathrm{MS}^{-1}(S_0)} c_S X_S \right).$$

We call each w_{S_0} a *pure summand* of w.

For $J \in \mathbb{N}^d$ and $S \in \mathrm{MS}^{-1}(S_0)$ define $X_S^J := [X_{s_1}^{(j_1)}, \ldots, X_{s_d}^{(j_d)}]$. We say that $(S, J) \sim (S', J')$ are equivalent if there exists $\sigma \in S_d$ such that $(\sigma S, \sigma J) = (S', J')$. We denote equivalence classes with respect to this action by $[(S, J)]$. Notice that X_S^J and $X_{S'}^{J'}$ have the same multiset of variables if and only if $(S, J) \sim (S', J')$. We therefore say that X_S^J is of type $[(S, J)]$.

Using Formula (5.1), by induction, we get the following technical lemma. Note that $\binom{u}{j_1, \ldots, j_d} = \frac{u!}{j_1! \ldots j_d!}$ are the multinomial coefficients counting the number of ways to put u distinct objects in d distinct bins, with j_k objects in the k-th bin.

LEMMA 5.3. *Let $S \in [r]^d$ and $w = X_S$. Then $w^{(u)} = \sum_{J \in \mathbb{N}^d : |J| = u} \binom{u}{j_1, \ldots, j_d} X_S^J$.*

COROLLARY 5.4. *Let $\mathfrak{g}$ be a simple K-Lie algebra, and let w be a d-homogeneous word of pure type S_0 such that $\varphi_w : \mathfrak{g}^r \to \mathfrak{g}$ is not trivial. Then for any $u \in \mathbb{N}$, $\tilde{J} \in \mathbb{N}^d$ with $|\tilde{J}| = u$, and any $\tilde{S} \in \mathrm{MS}^{-1}(S_0)$, the map $\varphi_w^{(u)}$ contains a non-trivial pure summand w' of type $[(\tilde{S}, \tilde{J})]$.*

PROOF. Write $w = \sum_{S \in \mathrm{MS}^{-1}(S_0)} c_S X_S$. Then by Lemma 5.3,

$$\varphi_w^{(u)} = \sum_{S \in \mathrm{MS}^{-1}(S_0)} \sum_{J:|J|=u} c_S \binom{u}{j_1, \dots, j_d} X_S^J.$$

In particular,

$$\varphi_{[(\tilde{S},\tilde{J})]} := \sum_{S \in \mathrm{MS}^{-1}(S_0)} \sum_{J:(S,J) \in [(\tilde{S},\tilde{J})]} c_S \binom{u}{j_1, \dots, j_d} X_S^J$$

is the pure summand of $\varphi_w^{(u)}$ of type $[(\tilde{S}, \tilde{J})]$. It is left to verify it is not trivial. Let W be the subspace of $J_m(\mathfrak{g}^r)$ defined by $X_s^{(u)} = X_s$ for all s and u. Note that

$$\left|\{J : (S, J) \in [(\tilde{S}, \tilde{J})]\}\right| = \frac{\left|\mathrm{Stab}_{S_d}(\tilde{S})\right|}{\left|\mathrm{Stab}_{S_d}((\tilde{S}, \tilde{J}))\right|} =: C([\tilde{S}, \tilde{J}]),$$

and hence there exists some $C'([\tilde{S}, \tilde{J}]) \in \mathbb{Z}_{>0}$ such that

$$\varphi_{[(\tilde{S},\tilde{J})]}|_W = \sum_{S \in \mathrm{MS}^{-1}(S_0)} C(\tilde{S}, \tilde{J}) c_S \binom{u}{j_1, \dots, j_d} X_S = C'([\tilde{S}, \tilde{J}]) \cdot \varphi_w,$$

which is non-trivial by our assumption. □

We next use the methods discussed in Subsection 5.1 to reduce $J_m(\varphi_w)$ to a simpler map.

PROPOSITION 5.5. *There exists a degeneration of $J_m(\varphi_w)$ into a generating map $\phi : \mathfrak{g}^{r(m+1)} \to \mathfrak{g}^{m+1}$ of the form $\phi = (\phi_0, \dots, \phi_m)$, where each $\phi_u : \mathfrak{g}^{r(m+1)} \to \mathfrak{g}$ is a Lie algebra word map. Furthermore there is a partition of the interval $[0, m]$ into intervals $\{[i_j, i_{j+1}]\}_j$ of length at most $2d + 1$, such that the collection $\{\psi_j\}_j$, where $\psi_j := (\phi_{i_j}, \phi_{i_j+1}, \dots, \phi_{i_{j+1}})$, has mutually disjoint variables. In particular, $\phi^{-1}(0) = \prod_j \psi_j^{-1}(0)$.*

PROOF. We choose an averaging weight

$$\omega_{\mathrm{av}}(X_s^{(u)}) := (d + 1)^u,$$

and observe that

$$\sigma_{\omega_{\mathrm{av}}}(J_m(\varphi_w)) = \left(\sigma_{\omega_{\mathrm{av}}}(\varphi_w), \sigma_{\omega_{\mathrm{av}}}(\varphi_w^{(1)}), \dots, \sigma_{\omega_{\mathrm{av}}}(\varphi_w^{(m)})\right),$$

where each $\sigma_{\omega_{\mathrm{av}}}(\varphi_w^{(u)})$ is a sum of pure summands of φ_w, and each pure summand is of type $[(S, J)]$, for some $J = (j_1, \dots, j_d)$ such that $\left|j_k - \frac{u}{d}\right| \leq 1$ for any $k \in [d]$. Note that each of these summands does not vanish by Corollary 5.4, and they cannot cancel each other as they consist of different monomials. We now order the variables $X_1, \dots, X_r, \dots, X_1^{(m)}, \dots, X_r^{(m)}$ from 1 to $r(m+1)$, by $\mathrm{ord}(X_s^{(u)}) = ur + s$, and set a monomialization weight

$$\omega_{\mathrm{mon}}(X_s^{(u)}) := (d + 1)^{\mathrm{ord}(X_s^{(u)})}.$$

Observe that for any $\sigma_{\omega_{\text{av}}}(\varphi_w^{(u)})$, there is a unique pure summand of type $[(S, J(u))]$, with $J(u) = (j_{u,1}, \dots, j_{u,d})$, such that $\deg_{\omega_{\text{mon}}}([X_{s_1}^{(j_{u,1})}, \dots, X_{s_d}^{(j_{u,d})}])$ is minimal. In particular, we have the following.

COROLLARY 5.6. *The map*

$$\widetilde{w}_u := \sigma_{\omega_{\text{mon}}}\left(\sigma_{\omega_{\text{av}}}(\varphi_w^{(u)})\right)$$

is a non-trivial d-homogeneous word map of pure type $[(S, J(u))]$ *with* $|J(u)| = u$ *and* $\left|j_{u,k} - \frac{u}{d}\right| \le 1$ *for any* $k \in [d]$.

Write $X^{(u)}$ (resp. $Y^{(u)}$) for the set of variables $X_1^{(u)}, \dots, X_r^{(u)}$ (resp. $Y_1^{(u)}, \dots, Y_r^{(u)}$). We now use the level separation method; we apply one convolution to $J_m(\varphi_w)$ to create a new set of $r(m+1)$ variables $Y, \dots, Y^{(m)}$, and set

$$\omega_{\text{LS}}(X^{(u)}) = \begin{cases} 0 & \text{if } u \bmod 4 = 0, 1 \\ d^2 + 1 & \text{if } u \bmod 4 = 2, 3, \end{cases}$$

$$\omega_{\text{LS}}(Y^{(u)}) = \begin{cases} d^2 & \text{if } u \bmod 4 = 0, 1 \\ 0 & \text{if } u \bmod 4 = 2, 3. \end{cases}$$

By Corollary 5.6 and by the definition of ω_{LS}, the map

$$\phi := \sigma_{\omega_{\text{LS}}}\left(\sigma_{\omega_{\text{mon}}}\left(\sigma_{\omega_{\text{av}}}(J_m(\varphi_w)^{*2})\right)\right)$$

is of the form $\phi = (\phi_0, \dots, \phi_m)$ where the dependence of ϕ_j on $\{X^{(u)}, Y^{(u)}\}_{u=0}^m$ is as follows:

(1) The polynomials $\phi_0, \dots, \phi_{\lceil \frac{3d}{2} \rceil - 1}$ depend only on the variables $X, X^{(1)}, X^{(2)}$.
(2) For every $j \in \mathbb{N}$, the polynomials $\phi_{\lceil \frac{(3+8j)d}{2} \rceil}, \dots, \phi_{\lfloor \frac{(7+8j)d}{2} \rfloor}$ depend only on the variables $Y^{(4j+1)}, Y^{(4j+2)}, Y^{(4j+3)}, Y^{(4j+4)}$.
(3) For every $j \in \mathbb{N}$, the polynomials $\phi_{\lfloor \frac{(7+8j)d}{2} \rfloor + 1}, \dots, \phi_{\lceil \frac{(11+8j)d}{2} \rceil - 1}$ depend only on the variables $X^{(4j+3)}, X^{(4j+4)}, X^{(4j+5)}, X^{(4j+6)}$.

Using the level separation method, we separated $\sigma_{\omega_{\text{mon}}}\left(\sigma_{\omega_{\text{av}}}(J_m(\varphi_w)^{*2}\right)$ into levels with pairwise disjoint variables, where each level is of size at most $(2d+1)\dim\mathfrak{g}$. Notice that $\phi_u = \widetilde{w}_u$ for any u, in either the variables $X, \dots, X^{(u)}$ or $Y, \dots, Y^{(u)}$, so every ϕ_u is non-trivial and thus generating. Since for each u the polynomial ϕ_u involves different monomials, it follows that the map $(\phi_{s_1}, \dots, \phi_{s_2})$ where

$$(s_1, s_2) \in \Big\{(0, \left\lceil \frac{3d}{2} \right\rceil - 1), (\left\lceil \frac{(3+8j)d}{2} \right\rceil, \left\lfloor \frac{(7+8j)d}{2} \right\rfloor), \\ (\left\lfloor \frac{(7+8j)d}{2} \right\rfloor + 1, \left\lceil \frac{(11+8j)d}{2} \right\rceil - 1)\Big\}$$

is generating as well. This finishes the proof of the proposition. $\square$

5.2. Lie algebras of low rank and small polyhypergraphs

In order to treat word maps on low rank Lie algebras and polyhypergraphs with few types of edges, we need the following Proposition 5.7. For a polynomial $f(x_1,\dots,x_n)=\sum_{I\in\mathbb{N}^n} a_I x^I$ set

$$d_{\mathrm{mon}}(f):=\max\{i_j : I=(i_1,\dots,i_n)\in\mathbb{N}^n \text{ and } a_I\neq 0\}.$$

For a polynomial map $\varphi=(\varphi_1,\dots,\varphi_m):\mathbb{A}_K^n\to\mathbb{A}_K^m$ set $d_{\mathrm{mon}}(\varphi):=\max_i\{d_{\mathrm{mon}}(\varphi_i)\}$.

PROPOSITION 5.7. *Let $\varphi=(\varphi_1,\dots,\varphi_m):\mathbb{A}_K^n\to\mathbb{A}_K^m$ be a generating morphism. Then $\varphi^{*(md_{\mathrm{mon}}(\varphi)+1)}$ is (FRS).*

PROOF. By [**GH19**, Section 6] we may assume that $K=\mathbb{Q}$. Write $\mu_{\mathbb{Z}_p}$ for the normalized Haar measure on $\mathbb{Z}_p$ and set $\tau_{p,\varphi}=\varphi_*(\mu_{\mathbb{Z}_p}^n)$, for any large enough prime p. By [**Clu04**, Theorem 6.1] and [**Chu76**], it follows that $\mathcal{F}(\tau_{p,f})\in L^{md_{\mathrm{mon}}(\varphi)+\epsilon}$ for every $\epsilon>0$. In particular, for any $t>md_{\mathrm{mon}}(\varphi)$ we have:

$$\|\mathcal{F}(\tau_{p,\varphi^{*t}})\|_1=\left\|\mathcal{F}(\tau_{p,\varphi}^{*t})\right\|_1=\left\|(\mathcal{F}(\tau_{p,\varphi}))^t\right\|_1=\|\mathcal{F}(\tau_{p,\varphi})\|_t^t<C,$$

so $\tau_{p,\varphi}^{*t}$ has continuous density. It now follows from [**GH19**, Proposition 3.16] that φ^{*t} is (FRS) (this is a consequence of an analytic criterion for the (FRS) property, see Theorem 8.5). □

Note that $d_{\mathrm{mon}}(\varphi_w)\leq d-1$ for any word map $\varphi_w:\mathfrak{g}^r\to\mathfrak{g}$ of degree $d\geq 2$. Corollaries 5.8 and 5.9 now follow directly from Proposition 5.7 and Lemma 3.8:

COROLLARY 5.8. *Let $w\in\mathcal{L}_r$ be a word of degree $d\in\mathbb{N}$ and let $\mathfrak{g}$ be a simple K-Lie algebra such that φ_w is non-trivial on $\mathfrak{g}$. Then,*

(1) *If $t\geq(\dim\mathfrak{g})^2$ then φ_w^{*t} is flat.*
(2) *If $t\geq d(\dim\mathfrak{g})$ then φ_w^{*t} is (FRS).*

Let $\Gamma:=(\mathcal{I},\mathcal{J},\mathcal{S})$ be a d-PH and let $(\Gamma,\mathrm{V},\{\tau_\gamma\})$ be an assignment, where $\mathrm{V}=(V(i))_{i\in\mathcal{I}}$ and $\{\tau_\gamma\}_{\gamma\in\mathcal{S}}$ is a collection of d-homogeneous multilinear forms

$$\tau_{\{i_1,\dots,i_d\},l}:\bigoplus_{i\in\{i_1,\dots,i_d\}}V(i)\to K.$$

Write

$$\Phi_{\Gamma,\mathrm{V},\{\tau_\gamma\}}:\bigoplus_{i\in\mathcal{I}}V(i)\to K^{\mathcal{J}}$$

for the corresponding map (as in (4.1)).

COROLLARY 5.9. *The t-th self-convolution $(\Gamma,\mathrm{V}^{\oplus t},\{\tau_\gamma^{\oplus t}\})$ of $(\Gamma,\mathrm{V},\{\tau_\gamma\})$ (Definition 4.34) is*

(1) *Flat if $t\geq|\mathcal{J}|^2$.*
(2) *(FRS) if $t\geq d|\mathcal{J}|+1$.*

Moreover, if Γ is a hypergraph consisting of a single hyperedge, with an assignment (Γ,V,τ), then (Γ,V,τ) is flat, and $(\Gamma,V^{\oplus t},\{\tau^{*t}\})$ is (FRS) if $t\geq d_{\mathrm{mon}}(\tau)+1$.

REMARK 5.10. Note that the map $\Phi_{\Gamma,\mathrm{V}^{\oplus t},\{\tau_\gamma^{*t}\}}$ can be naturally identified with $\Phi_{\Gamma,\mathrm{V},\{\tau_\gamma\}}^{*t}$.

5.3. Proof for $\mathfrak{sl}_n$

We assume that $n > d$, as otherwise we may use the results of Section 5.2. Let $\mathfrak{t}$ be the Cartan subalgebra of $\mathfrak{sl}_n$ consisting of diagonal matrices, write $\mathfrak{sl}_n = \mathfrak{t} \oplus \bigoplus_{1\leq i\neq j\leq n} \mathfrak{g}_{i,j}$ and consider the basis $\mathcal{B} = \{E_{i,j} : 1 \leq i \neq j \leq n\} \cup \{H_i\}_{i=1}^{n-1}$ with $0 \neq E_{i,j} \in \mathfrak{g}_{i,j}$ and $H_i = E_{i,i} - E_{i+1,i+1} \in \mathfrak{t}$. Set $e_{i,j} = E_{i,j}$ if $i \neq j$ and $e_{i,i} = H_i$ if $i = j < n$. We denote by $\Sigma(\mathfrak{sl}_n, \mathfrak{t})$ the corresponding root system, consisting of the roots $\{\epsilon_i - \epsilon_j\}_{i\neq j}$, where $\epsilon_i \in \mathfrak{t}^*$ is the projection map to the i-th coordinate.

Let $\mathcal{B}^* = \{\hat{e}_{i,j}\}$ be the dual basis of $\mathcal{B}$. Let w be a d-homogeneous word of pure type and consider the d-PH $\Gamma_{\mathfrak{sl}_n,w}$ and its assignment $(\Gamma_{\mathfrak{sl}_n,w}, V, \{\tau_\gamma\}_{\gamma\in\mathcal{S}})$ as defined in Section 4.5.2, for $\mathcal{I} = \mathcal{J} = [n]^2\backslash\{(n,n)\}$. Recall that $\Gamma_{\mathfrak{sl}_n,w}$ has vertices $\{e_{i,j}\}_{(i,j)\in\mathcal{I}}$ and hyperedges $\mathcal{S}$ of the form $(\{e_{i_1,j_1}, \ldots, e_{i_d,j_d}\}, \hat{e}_{k,l})$. Here, $(k,l) \in \mathcal{J}$ is the type of the hyperedge.

REMARK 5.11. Note that if $[e_{i_1,j_1}, \ldots, e_{i_d,j_d}] \in \mathfrak{g}_{k,l}$ with $k \neq l$, then for each permutation $\sigma \in S_d$, we have $[e_{i_{\sigma(1)},j_{\sigma(1)}}, \ldots, e_{i_{\sigma(d)},j_{\sigma(d)}}] \in \mathfrak{g}_{k,l}$. In particular, each edge $\{e_{i_1,j_1}, \ldots, e_{i_d,j_d}\}$ is of unique type $\hat{e}_{k,l}$. On the other hand, if $[e_{i_1,j_1}, \ldots, e_{i_d,j_d}] \in \mathfrak{t}$, the hyperedge $\{e_{i_1,j_1}, \ldots, e_{i_d,j_d}\}$ might be of several types. For example, take $w = [X_1, X_2]$, then

$$[e_{1,4}, e_{4,1}] = E_{1,1} - E_{4,4} = e_{1,1} + e_{2,2} + e_{3,3},$$

so the edge $\{e_{1,4}, e_{4,1}\}$ is of types $\hat{e}_{11}, \hat{e}_{22}$ and $\hat{e}_{33}$.

Our goal in this subsection is to prove the following theorem:

THEOREM 5.12. *Let $w \in \mathcal{L}_r$ be a d-homogeneous word of pure type and let $n > d$.*

(1) *The d-PH $(\Gamma_{\mathfrak{sl}_n,w}, V, \{\tau_\gamma\}_{\gamma\in\mathcal{S}})$ is flat if* $\dim V \geq 20d^3r$.
(2) *The d-PH $(\Gamma_{\mathfrak{sl}_n,w}, V, \{\tau_\gamma\}_{\gamma\in\mathcal{S}})$ is (FRS) if* $\dim V \geq 80d^3r$.

We would first like to reduce $\Gamma_{\mathfrak{sl}_n,w}$ to three simpler polyhypergraphs. Let $M = [3]$ and set a coloring function $\nu_0(e_{i,j}) = (2d(i-j), -1, 2d(j-i))$. Notice that given a hyperedge $(\{e_{i_1,j_1}, \ldots, e_{i_d,j_d}\}, \hat{e}_{k,l}) \in \mathcal{S}$,

$$(\star)\ \widetilde{\nu}_0(\{e_{i_1,j_1}, \ldots, e_{i_d,j_d}\}) = \sum_{u=1}^{d} \nu_0(e_{i_u,j_u}) = (2\sum_{u=1}^{d} d(i_u - j_u), -d, 2\sum_{u=1}^{d} d(j_u - i_u)).$$

Recall that in any simple Lie algebra $\mathfrak{g}$, for any $\alpha, \beta \in \Sigma(\mathfrak{g}, \mathfrak{t})$ we have $[\mathfrak{g}_\alpha, \mathfrak{g}_\beta] \subseteq \mathfrak{g}_{\alpha+\beta}$ if $\alpha + \beta \in \Sigma(\mathfrak{g}, \mathfrak{t})$, and zero otherwise. Since the hyperedge above is of type (k,l), it follows that $\sum_{u=1}^{d} \epsilon_{i_u} - \epsilon_{j_u} = \epsilon_k - \epsilon_l$. In particular, we get

$$(\star) = (2d(k-l), -d, 2d(l-k)),$$

so ν_0 is admissible, and hence by Corollary 4.32, we reduce to the following d-PHs:

(1) $\Gamma_+ := \mathrm{gr}_{\nu_0,1}(\Gamma_{\mathfrak{sl}_n,w})$ consisting of hyperedges of type $\hat{e}_{k,l}$ with $k < l$.
(2) $\Gamma_0 := \mathrm{gr}_{\nu_0,2}(\Gamma_{\mathfrak{sl}_n,w})$ consisting of hyperedges of type $\hat{e}_{k,k}$ with $k \in [n-1]$.
(3) $\Gamma_- := \mathrm{gr}_{\nu_0,3}(\Gamma_{\mathfrak{sl}_n,w})$ consisting of hyperedges of type $\hat{e}_{k,l}$ with $k > l$.

As we will see soon, the d-PHs Γ_+ and Γ_- can rather easily be reduced to hypergraphs. Γ_0 requires a special treatment, since by Remark 5.11, the hyperedges of Γ_0 may not have a unique type, so a non-careful enough use of the elimination method might result in a non-generating map (see Remark 4.31). We start by analyzing Γ_+ and Γ_-.

5.3.1. Reduction of Γ_+ and Γ_- to d-hypergraphs. We consider Γ_+, the proof for Γ_- is similar. Recall $\Gamma_+ := (\mathcal{I}, \mathcal{J}_+, \mathcal{S}_+)$ with $\mathcal{J}_+ = \mathrm{gr}_{\nu_0,1}(\mathcal{J}) = \{(k,l) \in \mathcal{J} : k < l\}$ and $\mathcal{S}_+ = \mathrm{gr}_{\nu_0,1}(\mathcal{S})$. The reduction to a hypergraph is done by applying the averaging method followed by the monomialization method.

Set a weight function $\omega_{\mathrm{av}}(e_{i,j}) = (d+1)^{|i-j|}$ for any i,j. Following the discussion in Subsection 5.1, ω_{av} leaves hyperedges $(\{e_{i_1,j_1}, \dots, e_{i_d,j_d}\}, \hat{e}_{k,l})$ such that each of $e_{i_1,j_1}, \dots, e_{i_d,j_d}$ has length $|i_m - j_m|$ which is as close as possible to the average, i.e. to $\frac{|k-l|}{d}$. Denote $\Gamma_1 = \mathrm{gr}_{\omega_{\mathrm{av}}}(\Gamma_+) = (\mathcal{I}, \mathcal{J}_+, \mathcal{S}_1)$, where

$$\mathcal{S}_1 := \{(\{e_{i_1,j_1}, \dots, e_{i_d,j_d}\}, \hat{e}_{k,l}) \in \mathcal{S}_+ \text{ such that } \widetilde{\omega}_{\mathrm{av}}(\{e_{i_1,j_1}, \dots, e_{i_d,j_d}\}) \text{ is minimal}\}.$$

Order $\{e_{i,j}\}$ by $\mathrm{ord}(e_{i,j}) = (i-1)n + j$ and set a monomialization weight

$$\omega_{\mathrm{mon}}(e_{i,j}) := (d+1)^{\mathrm{ord}(e_{i,j})}.$$

Note that for any $(k,l) \in \mathcal{J}_+$, there is a unique edge $(\{e_{i_1,j_1}, \dots, e_{i_d,j_d}\}, \hat{e}_{k,l}) \in \mathcal{S}_1$ with minimal weight $\widetilde{\omega}_{\mathrm{mon}}(\{e_{i_1,j_1}, \dots, e_{i_d,j_d}\})$. Indeed,

$$\begin{aligned} \widetilde{\omega}_{\mathrm{mon}}(\{e_{i_1,j_1}, \dots, e_{i_d,j_d}\}) &= \sum_{u=1}^{d} (d+1)^{\mathrm{ord}(e_{i_u,j_u})} \\ &= \sum_{p=1}^{n^2-1} (\#\{u : \mathrm{ord}(e_{i_u,j_u}) = p\}) \cdot (d+1)^p, \end{aligned} \tag{5.2}$$

so this decomposition is unique. Denote the resulting d-PH by $\Gamma_2 = \mathrm{gr}_{\omega_{\mathrm{mon}}}(\Gamma_1) := (\mathcal{I}, \mathcal{J}_+, \mathcal{S}_2)$.

PROPOSITION 5.13. *For any $(k,l) \in \mathcal{J}_+$ we have*

(1) *Γ_2 contains exactly one hyperedge $\{e_{i_1,j_1}, \dots, e_{i_d,j_d}\}$ of type (k,l) and it is of unique type.*
(2) *Any hyperedge $(\{e_{i_1,j_1}, \dots, e_{i_d,j_d}\}, \hat{e}_{k,l}) \in \mathcal{S}_2$ with $|k-l| \leq d$ consists of vertices e_{i_u,j_u} which are either simple roots $e_{i,i+1}$ or elements $e_{i,i}$ of the Cartan $\mathfrak{t}$.*

PROOF. Indeed, (5.2) and Remark 5.11 imply (1). It is left to prove (2). Assume $k < l \leq k+d$. Since $\mathcal{S}_2 \subseteq \mathcal{S}_1$, it is enough to show that any $(\{e_{i_1,j_1}, \dots, e_{i_d,j_d}\}, \hat{e}_{k,l}) \in \mathcal{S}_1$ satisfies (2), which is equivalent to

$$\widetilde{\omega}_{\mathrm{av}}(\{e_{i_1,j_1}, \dots, e_{i_d,j_d}\}) \leq (d+1)d. \tag{$\star$}$$

It is left to find a hyperedge of type $\hat{e}_{k,l}$ in Γ_+ that satisfies ($\star$). Without loss of generality assume that $k > \frac{d}{2}$ (a similar construction can be done for $k \leq \frac{d}{2}$). Further recall that $n > d$. Let $I_0 = ((i_{0,1}, j_{0,1}), \dots, (i_{0,d}, j_{0,d})) \in \mathcal{I}^d$ be the d-tuple defined by

$$\begin{cases} (i_{0,u}, j_{0,u}) := (k-u+1, k-u) & \text{if } u \leq \frac{d-(l-k)}{2} \\ (i_{0,u}, j_{0,u}) := (l+u-d-1, l+u-d) & \text{if } u > \frac{d-(l-k)}{2} \end{cases}$$

if $d-(l-k)$ is even, and as follows if $d-(l-k)$ is odd:

$$\begin{cases} (i_{0,u}, j_{0,u}) := (k-u+1, k-u) & \text{if } u < \frac{d-(l-k)+1}{2} \\ (i_{0,u}, j_{0,u}) := (k-u+1, k-u+1) & \text{if } u = \frac{d-(l-k)+1}{2} \\ (i_{0,u}, j_{0,u}) := (l+u-d-1, l+u-d) & \text{if } u > \frac{d-(l-k)+1}{2}. \end{cases}$$

In view of Remark 4.27, $\varphi_w : \mathfrak{sl}_n^r \to \mathfrak{sl}_n$ can be written as a restriction of a matrix word map $\varphi_{\widetilde{w}} : M_n^r \to M_n$, with $\widetilde{w} = \sum_{S\in \mathrm{MS}^{-1}(S_0)} d_S \cdot X_{s_1} \cdot \dots \cdot X_{s_d}$ for some $S_0 \in [r]^{(d)}$ and constants d_S. Write $\widetilde{I}_0 := \mathrm{MS}(I_0)$. By Lemma 4.26 we have

$$\tau_{\widetilde{I}_0,(k,l)}(X_1,\dots,X_r)$$
$$= \sum_{((i_1,j_1),\dots,(i_d,j_d))\in \mathrm{MS}^{-1}(\widetilde{I}_0)} \sum_{S\in \mathrm{MS}^{-1}(S_0)} d_S \prod_{u=1}^{d} a_{i_u,j_u,s_u,}\hat{e}_{k,l}\left(e_{i_1,j_1}\cdot \dots \cdot e_{i_d,j_d}\right),$$

where $X_s = \sum_{(i,j)\in\mathcal{I}} a_{i,j,s}e_{i,j}$. Now, note that $e_{(i_{0,1},j_{0,1})}\cdot \dots \cdot e_{(i_{0,d},j_{0,d})} \neq 0$, but any reordering of these terms is the zero matrix, namely $e_{(i_{0,\sigma(1)},j_{0,\sigma(1)})}\cdot\dots\cdot e_{(i_{0,\sigma(d)},j_{0,\sigma(d)})} = 0$ for any $1 \neq \sigma \in S_d$. We get,

$$\tau_{\widetilde{I}_0,(k,l)}(X_1,\dots,X_r) = \sum_{S\in \mathrm{MS}^{-1}(S_0)} d_S \prod_{u=1}^{d} a_{i_{0,u},j_{0,u},s_u}\hat{e}_{k,l}\left(e_{i_{0,1},j_{0,1}}\cdot \dots \cdot e_{i_{0,d},j_{0,d}}\right)$$
$$= \sum_{S\in \mathrm{MS}^{-1}(S_0)} d_S \prod_{u=1}^{d} a_{i_{0,u},j_{0,u},s_u}.$$

Note that the monomials

$$\{\prod_{u=1}^{d} a_{i_{0,u},j_{0,u},s_u}\}_{S\in \mathrm{MS}^{-1}(S_0)}$$

are mutually different (since the pairs $\{(i_{0,u},j_{0,u})\}_{u=1}^{d}$ are mutually different), and $d_S \neq 0$ for some $S \in \mathrm{MS}^{-1}(S_0)$, so $\tau_{\widetilde{I}_0,(k,l)}(X_1,\dots,X_r) \neq 0$, but by our construction of $\Gamma_{\mathfrak{sl}_n,w}$, we have $(\widetilde{I}_0, \hat{e}_{k,l}) \in \mathcal{S}$ and as it satisfies $(\star)$ we are done. □

Proposition 5.13 implies the following corollary.

COROLLARY 5.14. *The d-PH Γ_2 is induced from a d-hypergraph $\mathcal{G} = (\mathcal{V},\mathcal{E})$, with $\mathcal{V} = \mathcal{I}$ and*

$$\mathcal{E} = \{\text{the unique hyperedges } (\{e_{i_1,j_1},\dots,e_{i_d,j_d}\},\hat{e}_{k,l})$$
$$\in \mathcal{S}_2 \text{ of type } \hat{e}_{k,l}, \text{ for any } (k,l)\in \mathcal{J}_+\}.$$

Before we start to simplify $\mathcal{G}$ (or Γ_2) we would like to have a better understanding of its structure.

LEMMA 5.15. *Let $\Gamma_2 = (\mathcal{I},\mathcal{J}_+,\mathcal{S}_2)$ as above, and let k,l with $d < l-k$. Write $l-k = pd+q$, where $p,q\in\mathbb{N}$ and $0\le q\le d-1$. Then the hyperedge $(\{e_{i_1,j_1},\dots,e_{i_d,j_d}\},\hat{e}_{k,l}) \in \mathcal{S}_2$ satisfies $(i_u,j_u) = (k+(u-1)p, k+up)$ for any $1\le u\le d-q$, and*

$$(i_{d-q+g}, j_{d-q+g}) = (k+(d-q)p+(g-1)(p+1), k+(d-q)p+g(p+1)),$$

for $g\in\{1,\dots,q\}$.

PROOF. We first show that for $(i_1,j_1),\dots,(i_d,j_d)$ as in the lemma, we have $(\{e_{i_1,j_1},\dots,e_{i_d,j_d}\},\hat{e}_{k,l}) \in \mathcal{S}$. Indeed, since $i_1 = k$, $j_d = l$, $i_{u+1} = j_u$, and $\{i_u\}_{u=1}^{d} \cup \{j_d\}$ are mutually different, it follows that $e_{i_{\sigma(1)}j_{\sigma(1)}}\cdot\dots\cdot e_{i_{\sigma(d)}j_{\sigma(d)}} \neq 0$ if and only if $\sigma\in S_d$ is the trivial permutation. Using an argument similar to that in Proposition 5.13, we have $\tau_{\{(i_1,j_1),\dots,(i_d,j_d)\},(k,l)} \neq 0$. It is clear from the definitions of ω_{av} and

ω_{mon}, that $\widetilde{\omega}_{\mathrm{av}}(\{e_{i_1,j_1},\ldots,e_{i_d,j_d}\})$ is minimal over all hyperedges of type $\hat{e}_{k,l}$, and that $\widetilde{\omega}_{\mathrm{mon}}(\{e_{i_1,j_1},\ldots,e_{i_d,j_d}\})$ is minimal over all hyperedges in $\mathcal{S}_1$ of type $\hat{e}_{k,l}$. This implies that $(\{e_{i_1,j_1},\ldots,e_{i_d,j_d}\},\hat{e}_{k,l})\in\mathcal{S}_2$ as required. □

5.3.2. Proof for the d-hypergraph $\mathcal{G}=(\mathcal{V},\mathcal{E})$. We now reduce our hypergraph $\mathcal{G}=(\mathcal{V},\mathcal{E})$ to two simpler hypergraphs. We set $M=\{1,2\}$ and

$$\omega_2(e_{i,j})=\begin{cases}(d,0) & \text{if } i<j\\ (0,d^2) & \text{if } i\geq j.\end{cases}$$

This coloring results in two sub-hypergraphs:

(1) $\mathcal{G}_+=(\mathcal{V}_+,\mathcal{E}_+)$, consisting of vertices $\mathcal{V}_+=\{e_{i,j}\}_{i<j}$ corresponding to positive roots, and whose edges are

$$\mathcal{E}_+=\{(\{e_{i_1,j_1},\ldots,e_{i_d,j_d}\},\hat{e}_{k,l})\in\mathcal{E}: j_u>i_u\forall u\}=\{\hat{e}_{k,l}: l-k\geq d\}.$$

(2) $\mathcal{G}_{\mathrm{small}}=(\mathcal{V}_{\mathrm{small}},\mathcal{E}_{\mathrm{small}})$, consisting of $\mathcal{E}_{\mathrm{small}}=\mathcal{E}\backslash\mathcal{E}_+$. By Proposition 5.13 it follows that $\mathcal{V}_{\mathrm{small}}$ consists of vertices $e_{i,j}$ with $|i-j|\leq 1$.

We treat each of these hypergraphs separately.

Proof for $\mathcal{G}_+$:

The idea is to apply a series of level separations (see Subsection 5.1), to reduce $\mathcal{G}_+$ to the case of a hyperedge. We set the following coloring function $\omega_{\mathrm{LS},1}:\mathcal{V}_+\to\mathbb{N}^{\{0,1\}}$,

$$\omega_{\mathrm{LS},1}(e_{i,j})=\begin{cases}(0,2d) & \text{if } (j-i)\bmod 4=0,1\\ (2d+1,0) & \text{if } (j-i)\bmod 4=2,3.\end{cases}$$

Notice that an edge $\hat{e}_{k,l}\in\mathcal{E}_+$ (i.e. an edge where $l-k\geq d$)

(1) Has color 0 if and only if $|k-l|\bmod 4d\in\left[0,\left\lceil\frac{3d}{2}\right\rceil-1\right]\cup\left[\left\lfloor\frac{7d}{2}\right\rfloor+1,4d-1\right]$.
(2) Has color 1 if and only if $|k-l|\bmod 4d\in\left[\left\lceil\frac{3d}{2}\right\rceil,\left\lfloor\frac{7d}{2}\right\rfloor\right]$.

As a result, $\omega_{\mathrm{LS},1}$ splits the hypergraph $\mathcal{G}_+$ into levels (of size $\sim 2dn$):

(1) $\mathrm{gr}_{\omega_{\mathrm{LS},1},0}\mathcal{G}_+$ consisting of a disjoint union of hypergraphs

$$\mathcal{G}_{+,0,m}=(\mathcal{V}_{+,0,m},\mathcal{E}_{+,0,m}),$$

for $m\in\mathbb{N}$ such that

(a) $\mathcal{E}_{+,0,0}$ consists of edges $\hat{e}_{k,l}$ where $l-k\in\left[0,\left\lceil\frac{3d}{2}\right\rceil-1\right]$,
(b) $\mathcal{E}_{+,0,m}$ consists of edges $\hat{e}_{k,l}$ where

$$l-k\in\left[4d(m-1)+\left\lfloor\frac{7d}{2}\right\rfloor+1,4dm+\left\lceil\frac{3d}{2}\right\rceil-1\right].$$

(2) $\mathrm{gr}_{\omega_{\mathrm{LS},1},1}\mathcal{G}_+$ consisting of a disjoint union of hypergraphs

$$\mathcal{G}_{+,1,m}=(\mathcal{V}_{+,1,m},\mathcal{E}_{+,1,m}),$$

for $m\in\mathbb{N}$, where $\mathcal{E}_{+,1,m}$ consists of edges $\hat{e}_{k,l}$ where

$$l-k\in\left[4dm+\left\lceil\frac{3d}{2}\right\rceil,4dm+\left\lfloor\frac{7d}{2}\right\rfloor\right].$$

It is left now to deal with each of the hypergraphs $\mathcal{G}_{+,1,m}$ and $\mathcal{G}_{+,0,m}$. We prove our statement for $\mathcal{G}_{+,1,m}$, the proof for $\mathcal{G}_{+,0,m}$ is similar. Notice that any edge $\hat{e}_{k,l}$ consists of vertices $e_{i_1,j_1},\ldots,e_{i_d,j_d}\in\mathcal{V}_{+,1,m}$ such that j_u-i_u are as in Lemma 5.15. In particular, $j_u-i_u\in[4m+1,4m+4]$, implying the following:

LEMMA 5.16. *A sufficient condition for two edges $\hat{e}_{k,l}, \hat{e}_{k',l'} \in \mathcal{E}_{+,1,m}$ not to share a common vertex is $|k-k'| > (4m+4)d$.*

EXAMPLE 5.17. Take $n = 1000$, $d = 10$ and consider $\mathcal{G}_{+,1,20}$. Then $\mathcal{E}_{+,1,20}$ consists of the edges $\hat{e}_{k,l}$ such that $l-k \in [815, 835]$. Note that $\hat{e}_{1,821}$ shares common vertices with some of its neighbors such as $\hat{e}_{1,822}, \ldots, \hat{e}_{1,830}$ and $\hat{e}_{2,821}, \hat{e}_{3,821}, \ldots$. Apart from these close interactions, note that $\hat{e}_{1,821}$ and $\hat{e}_{83,903}$, which are far from each other, share the vertex $e_{83,165}$. Both types of interactions between edges will be taken into account.

We would therefore like to use level separation to minimize such interactions between edges. We start with taking care of far interactions. For any i let $\bar{i} = \lfloor \frac{i}{4m} \rfloor \bmod 2d$, and set $\omega_{\mathrm{LS},2} : \mathcal{V}_{+,1,m} \to \mathbb{N}^{\{0,\ldots,2d-1\}}$ as follows:

$$\omega_{\mathrm{LS},2}(e_{i,j})(s) := n + (n-j) \cdot \delta_{s\bar{j}} - (n-i) \cdot \delta_{s\bar{i}}.$$

Note that

$$\widetilde{\omega}_{\mathrm{LS},2}(\hat{e}_{k,l})(s) = dn + (n-l) \cdot \delta_{s\bar{l}} - (n-k) \cdot \delta_{s\bar{k}},$$

and in particular, the color of the edge $\hat{e}_{k,l}$ is always $\bar{k}$. As a consequence, for any $t \in \{0, \ldots, 2d-1\}$ the hypergraph $\mathrm{gr}_{\omega_{\mathrm{LS},2},t}(\mathcal{G}_{+,1,m})$ consists of edges $\hat{e}_{k,l}$ with $\bar{k} = t$. By Lemma 5.16, each $\mathrm{gr}_{\omega_{\mathrm{LS},2},t}(\mathcal{G}_{+,1,m})$ is a disjoint union of hypergraphs $\{\mathcal{G}^{t,e}_{+,1,m}\}_{e \in \mathbb{N}}$ with edges $\hat{e}_{k,l}$ where

$$k \in [4m(t+2ed), 4m(t+2ed) + 4m - 1].$$

Restricting to each of the smaller hypergraphs $\{G^{t,e}_{+,1,m}\}_{e \in \mathbb{N}}$, we obtain the following:

LEMMA 5.18. *A sufficient condition for two edges $\hat{e}_{k,l}, \hat{e}_{k',l'} \in \mathcal{E}^{t,e}_{+,1,m}$ not to share a common vertex is that either $|k-k'| > d$ or $|l-l'| > d$.*

Following Lemma 5.18, using two more level separations $\omega_{\mathrm{LS},3}$ and $\omega_{\mathrm{LS},4}$ (with d colors each), which are defined in a way similar to $\omega_{\mathrm{LS},2}$, only with $\bar{i} := i \bmod d$ instead of $\bar{i} := \lfloor \frac{i}{4m} \rfloor \bmod 2d$, we get that

(1) $\mathrm{gr}_{\omega_{\mathrm{LS},3},t_1} \mathcal{G}^{t,e}_{+,1,m}$ is a disjoint union of hypergraphs, each supported on the hyperedges $\hat{e}_{k_0,l_0}, \ldots, \hat{e}_{k_0,l_0+2d}$ with $l_0 = k_0 + 4dm + \lceil \frac{3d}{2} \rceil$, for a fixed k_0.
(2) $\mathrm{gr}_{\omega_{\mathrm{LS},4},t_2} \mathrm{gr}_{\omega_{\mathrm{LS},3},t_1} \mathcal{G}^{t,e}_{+,1,m}$ is a disjoint union of hypergraphs, each consisting of a single hyperedge $\hat{e}_{k_0,l_0}$.

Write $(\{e_{i_1,j_1}, \ldots, e_{i_d,j_d}\}, \hat{e}_{k_0,l_0})$ for any hyperedge of $\mathrm{gr}_{\omega_{\mathrm{LS},4},t_2} \mathrm{gr}_{\omega_{\mathrm{LS},3},t_1} \mathcal{G}^{t,e}_{+,1,u}$. Note that by Lemma 4.26, since each hyperedge of $\mathrm{gr}_{\omega_{\mathrm{LS},4},t_2} \mathrm{gr}_{\omega_{\mathrm{LS},3},t_1} \mathcal{G}^{t,e}_{+,1,u}$ is supported on distinct vertices, it follows that $\tau_{\{(i_1,j_1),\ldots,(i_d,j_d)\},(k_0,l_0)}$ is a sum of monomials of the form $x_1 x_2 \ldots x_d$. By Corollary 5.9, since $d_{\mathrm{mon}}(x_1 \cdot \cdots \cdot x_d) = 1$, it follows that $\mathrm{gr}_{\omega_{\mathrm{LS},4},t_2} \mathrm{gr}_{\omega_{\mathrm{LS},3},t_1} \mathcal{G}^{t,e}_{+,1,u}$ is flat if $\dim V \geq r$ and is (FRS) if $\dim V \geq 2r$.

Summary for $\mathcal{G}_+$ and $\mathcal{G}_-$:

We have used the level separation method four times, with coloring functions $\omega_{\mathrm{LS},1}, \omega_{\mathrm{LS},2}, \omega_{\mathrm{LS},3}$ and $\omega_{\mathrm{LS},4}$, with $2, 2d, d$ and d colors respectively, to reduce to a case of a hyperedge. For a hyperedge we need $\dim V \geq r$ to guarantee flatness, and $\dim V \geq 2r$ for the (FRS) property. Thus the hypergraph $\mathcal{G}_+$ (and similarly $\mathcal{G}_-$) is flat after $4d^3 = 2 \cdot 2d \cdot d \cdot d$ convolutions, and is (FRS) after $8d^3$ convolutions.

Proof for $\mathcal{G}_{\mathrm{small}}$:

Set $\bar{i} = i \bmod 2d$ and $\omega_{\mathrm{LS},5} : \mathcal{V}_{\mathrm{small}} \to \mathbb{N}^{\{0,\dots,2d-1\}}$ by

$$\omega_{\mathrm{LS},5}(e_{i,j})(s) := n + (n-j)\cdot\delta_{s\bar{j}} - (n-i)\cdot\delta_{s\bar{i}},$$

and note that $\hat{e}_{k.l}$ is of color $\bar{k}$ (recall we still have $l > k$).

By Proposition 5.13, since each of the vertices $e_{i_u.j_u}$ on which the hyperedge $\hat{e}_{k.l}$ is supported is either a simple root or an element of the Cartan, it follows that for each $s \in \{0,\dots,2d-1\}$, $\mathrm{gr}_{\omega_{\mathrm{LS},5},s}(\mathcal{G}_{\mathrm{small}})$ is a disjoint union of hypergraphs, each with at most d edges. Using another level separation $\omega_{\mathrm{LS},6}(e_{i,j})(s) := n-j\cdot\delta_{s\bar{j}}+i\cdot\delta_{s\bar{i}}$ of d colors (with $\bar{i} = i \bmod d$), we reduce to the case of a single hyperedge.

Summary for $\mathcal{G}_{\mathrm{small}}$: Since we used the coloring method twice, with $2d$ and d colors, by Corollary 5.9, we need $\dim V \geq 2d^2 r$ for $\mathcal{G}_{\mathrm{small}}$ to be flat, and $\dim V \geq 2d^3 r$ for $\mathcal{G}_{\mathrm{small}}$ to be (FRS).

We have therefore finished the cases of Γ_+ and Γ_-, so it is left to deal with Γ_0.

5.3.3. The case of Γ_0. Recall that $\Gamma_0 := \mathrm{gr}_{\nu_0,2}(\Gamma_{\mathfrak{sl}_n,w}) = (\mathcal{I}, \mathcal{J}_0, \mathcal{S}_0)$. In order to show that $(\Gamma_0, V, \{\tau_\gamma\}_{\gamma\in\mathcal{S}_0})$ is (FRS) for V sufficiently large, we can use the coloring and elimination methods, to reduce to a d-PH with few types of hyperedges, and then use Corollary 5.9. Instead, we prove that $\Phi_{\Gamma_0,V,\{\tau_\gamma\}}$ is jet-flat when V is large enough, using similar coloring and elimination techniques, only on the m-th jet map $J_m(\Phi_{\Gamma_0,V,\{\tau_\gamma\}})$. This approach will put us in a better position to treat the case of matrix word maps in Section 5.4.

Similarly to as done in Section 5.1.1, using an averaging weight, a monomialization weight, and the level separation method, we reduce to proving the flatness property of a map $\varphi_m = (\varphi_m^{(k_0)}, \dots, \varphi_m^{(k_0+2d)})$ whose target is a vector space of dimension $\mathrm{dim}\mathfrak{t}\cdot(2d+1) = (2d+1)(n-1)$, and each $\varphi_m^{(u)}$ is a certain pure summand of $\varphi_w^{(u)}$ (composed with $\{\hat{e}_{l,l}^{(u)}\}_{l\in[n-1]}$). The variables of this map are $\{e_{i,j}^{(u')}\}$ for $0 \leq u' \leq k_0 + 2d$ and $(i,j) \in [n]^2 \backslash \{(n,n)\}$. We now set the following weight

$$\nu_1(e_{i,j}^{(u')}) = \begin{cases} d & \text{if } |i-j| > 1 \text{ and } (i > d \text{ or } j > d) \\ 0 & \text{if } |i-j| \leq 1 \text{ or } i,j \leq d. \end{cases}$$

LEMMA 5.19. *The map $\mathrm{gr}_{\nu_1}(\varphi_m)$ is a generating polynomial map on variables $\{e_{i,j}^{(u')}\}$ with either $i,j \leq d$ or $|i-j| \leq 1$.*

PROOF. Note that

$$\mathrm{gr}_{\nu_1}(\varphi_m) = \left(\mathrm{gr}_{\nu_1}\varphi_m^{(k_0)}, \dots, \mathrm{gr}_{\nu_1}\varphi_m^{(k_0+2d)}\right).$$

Since $\varphi_m^{(u_1)}$ and $\varphi_m^{(u_2)}$ involve distinct monomials if $u_1 \neq u_2$, the maps $\mathrm{gr}_{\nu_1}\varphi_m^{(u_1)}$ and $\mathrm{gr}_{\nu_1}\varphi_m^{(u_2)}$ involve distinct monomials as well. Therefore, $\mathrm{gr}_{\nu_1}(\varphi_m)$ is generating if and only if $\mathrm{gr}_{\nu_1}\varphi_m^{(u)}$ is generating for every $k_0 \leq u \leq k_0 + 2d$. Let W be the subspace of $J_m(\mathfrak{sl}_n^r)$ defined by the relations $X_s^{(u')} = X_s$ for every u' and $s \in [r]$. By Corollary 5.4, $\varphi_m^{(u)}|_W = \lambda \cdot \varphi_w^{(0)}$ for some $\lambda > 0$. Thus we may assume that $u = 0$ and it is enough to show that $\Phi_{(\mathrm{gr}_{\nu_1}\Gamma_0,V,\{\tau_\gamma\}_{\gamma\in\mathcal{S}_0})}$ is generating.

Consider the Lie subalgebra $\mathfrak{h} \subseteq \mathfrak{sl}_n$ generated by the simple roots $e_{1,2}, \dots, e_{d-1,d}$. Since $\mathfrak{h} \simeq \mathfrak{sl}_d$, Corollary 4.8 implies that $\mathrm{gr}_{\nu_1}\varphi_w|_{\mathfrak{h}} = \varphi_w|_{\mathfrak{h}} \neq 0$, and Proposition 4.5 implies that $\Phi_{(\mathrm{gr}_{\nu_1}\Gamma_0,V,\{\tau_\gamma\}_{\gamma\in\mathcal{S}_0})}$ is generating on the types $\{\hat{e}_{l,l}\}_{l\in[d-1]}$ using only the variables of $\mathfrak{h}$.

We use a similar construction as in the proof of Proposition 5.13(2), only now for hyperedges of type $\hat{e}_{k,k}$. For simplicity assume that d is even (the case of an odd d is similar). For any $d+1 \leq k \leq n$, consider the d-tuple $I_0(k) = ((i_{0,1}, j_{0,1}), \dots, (i_{0,d}, j_{0,d})) \in \mathcal{I}^d$ defined by

$$\begin{cases} (i_{0,c}, j_{0,c}) := (k-c+1, k-c) & \text{if } 1 \leq c \leq \frac{d}{2} \\ (i_{0,c}, j_{0,c}) := (k+c-d-1, k+c-d) & \text{if } \frac{d}{2} < c \leq d, \end{cases}$$

and let $\widetilde{I_0}(k) = \mathrm{MS}(I_0(k))$. Note that here, unlike in the case of Γ_+ and Γ_-, the hyperedge $\widetilde{I_0}(k)$ is supported on several types, i.e. on the types $\{\hat{e}_{l,l}\}_{l \in [k-\frac{d}{2}, k-1]} \in \mathcal{J}_0$. By a similar computation as in Proposition 5.13, and by induction on k, it can be shown that the forms $\{\tau_\gamma\}$ supported on $\mathfrak{h}$ and on the hyperedges $\{\widetilde{I_0}(l)\}_{l \leq k}$ are generating on the types $\{\hat{e}_{l,l}\}_{l \in [k-1]}$, so we are done. $\square$

Back to our morphism $\mathrm{gr}_{\nu_1}(\varphi_m)$. Our choice of ν_1 guarantees that $\hat{e}_{k,k}^{(u)} \circ \mathrm{gr}_{\nu_1}(\varphi_m)$ is supported on variables $e_{i,j}^{(u')}$ with $|i-k| \leq 2d$. We set the following (monomialization) weight

$$\nu_2(e_{i,j}^{(u')}) = \begin{cases} (d+1)^i & \text{if } i > 2d \\ 0 & \text{if } i \leq 2d. \end{cases}$$

Similarly to the proof of Lemma 5.19, it can be shown that $\mathrm{gr}_{\nu_2}(\mathrm{gr}_{\nu_1}(\varphi_m))$ is generating. We now use the level separation method. Set $\bar{i} = \lfloor \frac{i}{4d} \rfloor \bmod 2$, and define a coloring function ν_{LS} with two colors $s \in \{0,1\}$ by

$$\nu_{\mathrm{LS}}(e_{i,j}^{(u)})(s) = (d^2+1) \cdot \delta_{s\bar{i}} + d^2 \cdot \delta_{s(1-\bar{i})}.$$

In view of Remark 4.33, our choice of ν_2 makes ν_{LS} an admissible coloring for the map $\mathrm{gr}_{\nu_2}(\mathrm{gr}_{\nu_1}(\varphi_m))$. Then each of $\{\mathrm{gr}_{\nu_{\mathrm{LS}},s}(\mathrm{gr}_{\nu_2}(\mathrm{gr}_{\nu_1}(\varphi_m)))\}_{s=1}^2$ is a collection of morphisms in distinct variables into affine spaces of dimension at most $6d \cdot (2d+1) \leq 15d^2$. By Corollary 5.7, the map $\mathrm{gr}_{\nu_{\mathrm{LS}},s}(\mathrm{gr}_{\nu_2}(\mathrm{gr}_{\nu_1}(\varphi_m)))^{*15d^3}$ is normal (as $d_{\mathrm{mon}} \leq d-1$). By Corollaries 2.15 and 3.12, and since we have used four colors, the assignment $(\Gamma_0, V, \{\tau_\gamma\}_{\gamma \in \mathcal{S}_0})$ is (FRS) if $\dim V \geq 60d^3 r = 4 \cdot 15d^3 r$. In particular, following the same proof for $m=0$, we reduce to a disjoint set of morphisms into a $6d$-dimensional vector space, so $6d^2$ convolutions are enough to ensure each map is flat, and $(\Gamma_0, V, \{\tau_\gamma\}_{\gamma \in \mathcal{S}_0})$ is flat if $\dim V \geq 12d^2 r$.

5.3.4. Proof of Theorem 5.12. Assume $n > d \geq 2$. We have seen that the hypergraphs $\mathcal{G}_+$ and $\mathcal{G}_-$ are flat after $4d^3$ convolutions, and (FRS) after $8d^3$ convolutions. $\mathcal{G}_{\mathrm{small}}$ is flat after $2d^2$ convolutions and (FRS) after $2d^3$ convolutions. Therefore, each of the d-PHs Γ_+ and Γ_- are flat after $5d^3$ convolution and (FRS) after $10d^3$ convolutions. We have seen that Γ_0 is flat after $12d^2$ and (FRS) after $60d^3$ convolutions. Theorem 5.12 now follows by combining the results for Γ_+, Γ_- and Γ_0 and using Corollary 4.32 repeatedly.

5.4. Matrix word maps

The proof for matrix word maps (Theorem 4.10) is similar to the $\mathfrak{sl}_n$ case. With the notations of Section 5.3, let

$$e_{\mathrm{cent}} = H_{\mathrm{cent}} = \sum_{i=1}^{n} \frac{E_{i,i}}{n}$$

and $\hat{e}_{\text{cent}} = \text{Tr}(\cdot)$ be such that $\{\hat{e}_{i,j}\} \cup \{\hat{e}_{\text{cent}}\}$ is the basis dual to $\{e_{i,j}\} \cup \{e_{\text{cent}}\}$. Let $w \in \mathcal{A}_r$ be a matrix word of degree d. We identify φ_w with a map $(\widetilde{\varphi}_w, \text{Tr}(\varphi_w)) : M_n^r \to \mathbb{A}^{n^2}$, where $(\widetilde{\varphi}_w)_{i,j} = \hat{e}_{i,j} \circ \varphi_w$ and $\text{Tr}(\varphi_w) = \hat{e}_{\text{cent}} \circ \varphi_w$. Using Lemma 4.19, we degenerate $(\widetilde{\varphi}_w, \text{Tr}(\varphi_w))$ to its highest degree homogeneous part $\psi_{w_0,w_1} := (\widetilde{\varphi}_{w_0}, \text{Tr}(\varphi_{w_1}))$, so that φ_{w_1} (resp. φ_{w_0}) is the highest degree homogeneous part of φ_w whose trace is non-zero (resp., $\widetilde{\varphi}_{w_0}$ is non-zero). Since the image of ψ_{w_0,w_1} is conjugate invariant, it must generate either M_n or $\mathfrak{sl}_n$, or the center of M_n. However, since both $\widetilde{\varphi}_{w_0}$ and $\text{Tr}(\varphi_{w_1})$ are non-zero, ψ_{w_0,w_1} is generating, even though it is not necessarily a matrix word map.

EXAMPLE 5.20. Let $\varphi_w(X,Y) = X^3 + [[[X,Y],Y],Y]$. Then $\varphi_{w_0} = [[[X,Y],Y],Y]$ and $\varphi_{w_1} = X^3$.

DEFINITION 5.21. A d-homogeneous word $w \in \mathcal{A}_r$ is said to be of *pure type* $S_0 \in [r]^{(d)}$ if it is of the form

$$w = \sum_{S \in \text{MS}^{-1}(S_0)} c_S X_{s_1} \cdot X_{s_2} \cdot \ldots \cdot X_{s_d}.$$

LEMMA 5.22. *We may assume that the words w_0 and w_1 appearing in $\psi_{w_0,w_1} = (\widetilde{\varphi}_{w_0}, \text{Tr}(\varphi_{w_1}))$ are homogeneous matrix words of pure type.*

PROOF. We use the elimination method similarly as in Proposition 4.18. We set a monomialization weight $\omega_{\text{mon}}(X_i) = (d+1)^i$ and then $\text{gr}_{\omega_{\text{mon}}}(\psi_{w_0,w_1})$ is of the required form. $\square$

PROOF OF THEOREM 4.10. If $n \leq d$, then Proposition 5.7 implies that for any $w \in \mathcal{A}_r$ with $\varphi_w : M_n^r \to M_n$ generating, the map $\varphi_w^{*d^3+1}$ is (FRS).

Now assume that $n > d$. We use an idea similar to the one used in the reduction to the d-PHs Γ_+, Γ_- and Γ_0 in the case of $\mathfrak{sl}_n$. We use the coloring function $\nu_0(e_{i,j}) = (2d(i-j), -1, 2d(j-i))$ for the polynomial map ψ_{w_0,w_1} (see discussion after Theorem 5.12). We therefore reduce to a disjoint union of three morphisms

$$\Psi_+ := (\hat{e}_{k,l})_{k<l} \circ (\widetilde{\varphi}_{w_0}) : M_n^r \to \mathbb{A}^{\frac{n(n-1)}{2}},$$
$$\Psi_- := (\hat{e}_{k,l})_{k>l} \circ (\widetilde{\varphi}_{w_0}) : M_n^r \to \mathbb{A}^{\frac{n(n-1)}{2}},$$
$$\Psi_0 := ((\hat{e}_{1,1}, \ldots, \hat{e}_{n-1,n-1}) \circ \widetilde{\varphi}_{w_0}, \text{Tr}(\varphi_{w_1})) : M_n^r \to \mathbb{A}^n.$$

Ψ_+ and Ψ_- can be encoded in terms of assignments of d-PHs named $\widetilde{\Gamma}_+$ and $\widetilde{\Gamma}_-$ as in the $\mathfrak{sl}_n$ case, only now the forms in play will encode matrix word maps instead of Lie algebra word maps. The proof for $\widetilde{\Gamma}_+$ and $\widetilde{\Gamma}_-$ is identical to the proof for Γ_+ and Γ_- in the $\mathfrak{sl}_n$ case. In particular, the maps Ψ_+^{*t} and Ψ_-^{*t} are flat if $t \geq 5d^3$ and (FRS) if $t \geq 10d^3$.

The proof for Ψ_0 is done by first dealing with $\psi_{w_0} := (\hat{e}_{1,1}, \ldots, \hat{e}_{n-1,n-1}) \circ \widetilde{\varphi}_{w_0}$, similarly as in Γ_0 for $\mathfrak{sl}_n$, so that $\psi_{w_0}^{*12d^2}$ is flat and $\psi_{w_0}^{*60d^3}$ is (FRS). In particular, we get that the dimensions of the fibers of $\Psi_0^{*12d^2}$ are at most $12d^2rn^2 - n + 1$, so by Lemma 3.9, $\Psi_0^{*24d^2}$ is flat. Using a similar argument (Lemma 3.9) on a certain degeneration of the m-th jets of Ψ_0 (as in the case of Γ_0 for $\mathfrak{sl}_n$), we deduce that $\Psi_0^{*200d^4}$ is (FRS).

We therefore get that φ_w^{*t} is flat if $t \geq 30d^3 > 10d^3 + 24d^2$ and is (FRS) if $t \geq 300d^4$. Proposition 4.21 allows us to reduce to the case of self-convolutions of matrix word maps (recall that when doing this the required number of convolutions is multiplied by 2). Combining this with the above arguments concludes the proof. $\square$

5.5. Proof of the general case

Let $\mathfrak{g}$ be a simple K-Lie algebra. We may assume that K is algebraically closed. Since we already dealt with Lie algebras of small rank we may assume that $\mathfrak{g}$ is either $\mathfrak{so}_{2n+1}, \mathfrak{sp}_{2n}$ or $\mathfrak{so}_{2n}$, and $n > 4d$. Write

$$\mathfrak{g} = \mathfrak{t} \oplus \bigoplus_{\alpha \in \Sigma(\mathfrak{g},\mathfrak{t})} \mathfrak{g}_\alpha$$

with $\mathfrak{t}$ a Cartan subalgebra, $\Sigma(\mathfrak{g},\mathfrak{t})$ the corresponding root system and $\triangle$ a basis of $\Sigma(\mathfrak{g},\mathfrak{t})$. Choose a Chevalley basis $\{h_\alpha\}_{\alpha\in\triangle} \cup \{e_\alpha\}_{\alpha\in\Sigma(\mathfrak{g},\mathfrak{t})}$. We define the *length* $\ell(\alpha)$ of a root $\alpha \in \Sigma^+(\mathfrak{g},\mathfrak{t})$ to be the number of simple roots $\alpha_{i_1}, \ldots, \alpha_{i_{\ell(\alpha)}}$ such that

$$\alpha = \sum_{j=1}^{\ell(\alpha)} \alpha_{i_j}.$$

We index this basis and its dual as $\{e_i\}_{i\in\mathcal{I}}$ and $\{\widehat{e}_j\}_{j\in\mathcal{J}}$ respectively. Let $w \in \mathcal{L}_r$ be a d-homogeneous word of pure type. We consider the assignment $(\Gamma_{\mathfrak{g},w}, V, \{\tau_\gamma\}_{\gamma\in\mathcal{S}})$ of the d-PH $\Gamma_{\mathfrak{g},w} = (\mathcal{I},\mathcal{J},\mathcal{S})$. Recall that $\Gamma_{\mathfrak{g},w}$ has vertices $\{e_i\}_{i\in\mathcal{I}}$ and hyperedges of the form $(\{e_{i_1},\ldots,e_{i_d}\},\widehat{e}_j)$. Such an edge has type j. We would like to prove the following:

THEOREM 5.23. *Let $w \in \mathcal{L}_r$ be a d-homogeneous word of pure type and let $n > 4d$. Then the d-PH $(\Gamma_{\mathfrak{g},w}, V, \{\tau_\gamma\}_{\gamma\in\mathcal{S}})$ is flat if* $\dim V \geq 4\cdot 10^4 d^3 r$, *and (FRS) if* $\dim V \geq 8 \cdot 10^4 d^4 r$.

As in the case of $\mathfrak{sl}_n$, in order to split $\Gamma_{\mathfrak{g},w}$ into simpler d-PHs we use the coloring function

$$\nu_0(\alpha) = \begin{cases} (-2d\cdot\ell(\alpha), -1, 2d\cdot\ell(\alpha)) & \text{if } \alpha\in\Sigma^+(\mathfrak{g},\mathfrak{t}) \\ (2d\cdot\ell(\alpha), -1, -2d\cdot\ell(\alpha)) & \text{if } \alpha\in\Sigma^-(\mathfrak{g},\mathfrak{t}) \\ (0,-1,0) & \text{else.} \end{cases}$$

We get three simpler d-PHs:

(1) $\Gamma_+ := \mathrm{gr}_{\nu_0,1}(\Gamma_{\mathfrak{g},w}) = (\mathcal{I},\mathcal{J}_+,\mathcal{S}_+)$ consisting of hyperedges of type $\widehat{e}_\delta$ with $\delta\in\Sigma^+(\mathfrak{g},\mathfrak{t})$.
(2) $\Gamma_0 := \mathrm{gr}_{\nu_0,2}(\Gamma_{\mathfrak{g},w}) = (\mathcal{I},\mathcal{J}_0,\mathcal{S}_0)$ consisting of hyperedges of type $\widehat{h}_\alpha$ with $\alpha\in\triangle$.
(3) $\Gamma_- := \mathrm{gr}_{\nu_0,3}(\Gamma_{\mathfrak{g},w}) = (\mathcal{I},\mathcal{J}_-,\mathcal{S}_-)$ consisting of hyperedges of type $\widehat{e}_\delta$ with $\delta\in\Sigma^-(\mathfrak{g},\mathfrak{t})$.

5.5.1. Reduction of Γ_+ and Γ_- to d-hypergraphs. Let $\mathfrak{g}$ be either $\mathfrak{so}_{2n}$, $\mathfrak{so}_{2n+1}$ or $\mathfrak{sp}_{2n}$ and consider the following bilinear forms:

$$\mathrm{B}_{2n} := \begin{pmatrix} 0 & I_n \\ I_n & 0 \end{pmatrix}, \ \mathrm{B}_{2n+1} := \begin{pmatrix} 0 & I_n & 0 \\ I_n & 0 & 0 \\ 0 & 0 & 1 \end{pmatrix} \ \text{ and } \ \mathrm{J}_n := \begin{pmatrix} 0 & I_n \\ -I_n & 0 \end{pmatrix}.$$

These forms induce embeddings $\mathfrak{so}_{2n} \hookrightarrow \mathfrak{so}_{2n+1} \hookrightarrow \mathfrak{gl}_{2n+1}$ and $\mathfrak{sp}_{2n} \hookrightarrow \mathfrak{gl}_{2n}$ such that $\mathrm{diag}_{2n+1} \cap \mathfrak{g}$ is a Cartan subalgebra of $\mathfrak{g}$, where $\mathrm{diag}_{2n+1} \subseteq \mathfrak{gl}_{2n+1}$ is the space of diagonal matrices. Let $\epsilon_i \in \mathfrak{t}^*$ be the restriction of the projection to diag_n to its i-th coordinate.

DEFINITION 5.24. Let $\mathfrak{g}$ be a simple classical Lie algebra. A root $\alpha \in \Sigma(\mathfrak{g}, \mathfrak{t})$ is said to be of

(1) Type a if it is of the form $a_{i,j} := \epsilon_i - \epsilon_j$, for $i \neq j$.
(2) Type b if it is of the form $b_{i,j} := \epsilon_i + \epsilon_j$, for any i, j (including $2\epsilon_i$ in the case of $\mathfrak{sp}_{2n}$).
(3) Type c if it is of the form $c_i := \epsilon_i$.

Notice that $\mathfrak{sl}_n$ contains only roots of type a, $\mathfrak{so}_{2n}$ and $\mathfrak{sp}_{2n}$ contain only roots of type a and b and $\mathfrak{so}_{2n+1}$ contains roots of type a, b and c.

LEMMA 5.25. *Let $\delta \in \Sigma^+(\mathfrak{g}, \mathfrak{t})$ and assume that there exist $\alpha_{i_1}, \ldots, \alpha_{i_d} \in \Sigma^+(\mathfrak{g}, \mathfrak{t})$ such that $\widehat{e}_\delta([e_{\alpha_{i_1}}, \ldots, e_{\alpha_{i_d}}]) \neq 0$. Then*

(1) *δ is of type a if and only if all $\{\alpha_{i_j}\}$ are of type a.*
(2) *δ is of type b if and only if either,*
    (a) *exactly $d-1$ of $\{\alpha_{i_j}\}$ are of type a, and α_{i_k} is of type b for a single $k \in [d]$, or;*
    (b) *exactly $d-2$ of $\{\alpha_{i_j}\}$ are of type a, and $\alpha_{i_{k_1}}$ and $\alpha_{i_{k_2}}$ are of type c for $k_1, k_2 \in [d]$.*
(3) *δ is of type c if and only if exactly $d-1$ of $\{\alpha_{i_j}\}$ are of type a, and for a single $k \in [d]$ the root α_{i_k} is of type c.*

We set an averaging weight (as in Section 5.3.1)

$$\omega_{\mathrm{av}}(e_\alpha) = (d+1)^{\ell(\alpha)} \text{ and } \omega_{\mathrm{av}}(h_\alpha) = 1,$$

and denote by $\Gamma_1 = \mathrm{gr}_{\omega_{\mathrm{av}}}(\Gamma_+) = (\mathcal{I}, \mathcal{J}_+, \mathcal{S}_1)$. Order $\{e_i\}_{i \in \mathcal{I}}$ as follows; choose an arbitrary order $\mathrm{ord}(h_\alpha) \in [n]$ for $\{h_\alpha\}_{\alpha \in \triangle}$, and,

$$\mathrm{ord}(e_\alpha) := \begin{cases} n + (i-1)n + j & \text{if } \alpha = a_{i,j} \\ n + 4n^2 + (i-1)n + j & \text{if } \alpha = b_{i,j} \\ n + 8n^2 + (i-1)n & \text{if } \alpha = c_i. \end{cases}$$

Set a monomialization weight

$$\omega_{\mathrm{mon}}(e_i) := (d+1)^{\mathrm{ord}(e_i)},$$

and denote the resulting d-PH by $\Gamma_2 = \mathrm{gr}_{\omega_{\mathrm{mon}}}(\Gamma_1) = (\mathcal{I}, \mathcal{J}_+, \mathcal{S}_2)$.

PROPOSITION 5.26. *Let $\delta \in \Sigma^+(\mathfrak{g}, \mathfrak{t})$ be a positive root. Then we have the following:*

(1) *Γ_2 contains exactly one hyperedge $\{e_{i_1}, \ldots, e_{i_d}\}$ of type $\widehat{e}_\delta$, and it is unique of this type.*
(2) *If $\ell(\delta) \leq 4d$, then the edge $(\{e_{i_1}, \ldots, e_{i_d}\}, \widehat{e}_\delta) \in \mathcal{S}_2$ consists of vertices $e_{i_j} = e_{\alpha_{i_j}}$ with $\ell(\alpha_{i_j}) \leq 8d$ or $e_{\alpha_{i_j}} = h_{\alpha_{i_j}}$.*

PROOF. The proof of (1) is similar to (1) of Proposition 5.13. To prove Item (2), it is enough by the definition of the weight ω_{av}, to prove the existence of $(\{e_{i_1}, \ldots, e_{i_d}\}, \hat{e}_\delta) \in \mathcal{S}_+$ with $e_{i_j} = e_{\alpha_{i_j}}$ and $\ell(\alpha_{i_j}) \leq 8d$, or $e_{i_j} = h_{\alpha_{i_j}}$. Let $\beta_1, \ldots, \beta_t$ be the simple roots appearing when writing δ as a sum of simple roots. Complete it to any connected subdiagram that consists of $4d$ simple roots (recall our assumption that $\mathrm{rk}(\mathfrak{g}) > 4d$). Such a subdiagram is isomorphic to a Dynkin diagram of some classical simple Lie subalgebra $\mathfrak{h}$ of rank $4d$. By Corollary 4.8 and Proposition 4.5, $\varphi_w|_{\mathfrak{h}^r} : \mathfrak{h}^r \to \mathfrak{h}$ is generating, so we may find $(\{e_{i_1}, \ldots, e_{i_d}\}, \hat{e}_\delta) \in \mathcal{S}_+$ with $e_{i_j} \in \mathfrak{h}$. Since $\mathrm{rk}\mathfrak{h} = 4d$ it follows that $\ell(e_{i_j}) \leq 8d$, so we are done. □

Proposition 5.26 implies the following:

COROLLARY 5.27. *The d-PH Γ_2 is induced from a d-hypergraph $\mathcal{G} = (\mathcal{V}, \mathcal{E})$, with $\mathcal{V} = \mathcal{I}$ and*

$$\mathcal{E} = \{\textit{the unique hyperedges } \{e_{i_1}, \ldots, e_{i_d}\} \textit{ in } \Gamma_2 \textit{ of type } \widehat{e}_\delta, \textit{ for any } \delta \in \mathcal{J}_+\}.$$

We would now like to reduce $\mathcal{G}$ (or Γ_2) into simpler hypergraphs. For this we need a more general analogue of Lemma 5.15.

LEMMA 5.28. *Let $\Gamma_2 = (\mathcal{I}, \mathcal{J}_+, \mathcal{S}_2)$ be as above, and let $\hat{e}_\delta \in \mathcal{J}_+$ be such that $\ell(\delta) \geq 4d$. Then there exist $\alpha_{i_1}, \alpha_{i_2}, \ldots, \alpha_{i_d} \in \Sigma^+(\mathfrak{g}, \mathfrak{t})$ such that*

$$(\{e_{\alpha_{i_1}}, e_{\alpha_{i_2}}, \ldots, e_{\alpha_{i_d}}\}, \widehat{e}_\delta) \in \mathcal{S}_2$$

and

$$\ell(\alpha_{i_j}) \in [\left\lfloor \frac{\ell(\delta)}{d} \right\rfloor - 1, \left\lceil \frac{\ell(\delta)}{d} \right\rceil + 1]$$

for every $j \in [d]$.

PROOF. Let $\delta \in \Sigma^+(\mathfrak{g}, \mathfrak{t})$. If δ is of type a, this is just Lemma 5.15. If δ is of type b, we may always find a tuple $(\alpha_{i_1}, \ldots, \alpha_{i_d})$ of positive roots $\alpha_{i_j} = \epsilon_{k_j} \pm \epsilon_{k_{j+1}}$ where $d - 1$ roots are of type a, and one root is of type b, whose lengths are in the interval

$$[\left\lfloor \frac{\ell(\delta)}{d} \right\rfloor - 1, \left\lceil \frac{\ell(\delta)}{d} \right\rceil + 1],$$

such that $\hat{e}_\delta([e_{\alpha_{i_1}}, \ldots, e_{\alpha_{i_d}}]) \neq 0$ and such that the $\{k_j\}_{j \in [d]}$ are distinct (the flexible choice of their lengths allows us to make sure that the $\{k_j\}$ are indeed distinct). A similar argument as in the proof of Proposition 5.13 gives $(\{e_{\alpha_{i_1}}, \ldots, e_{\alpha_{i_d}}\}, \widehat{e}_\delta) \in \mathcal{S}_+$. By definition of ω_{av}, any hyperedge $(\{e_{\beta_{j_1}}, \ldots, e_{\beta_{i_d}}\}, \widehat{e}_\delta) \in \mathcal{S}_1$ must satisfy that

$$\ell(\beta_{j_u}) \in [\left\lfloor \frac{\ell(\delta)}{d} \right\rfloor - 1, \left\lceil \frac{\ell(\delta)}{d} \right\rceil + 1].$$

Since $\mathcal{S}_2 \subseteq \mathcal{S}_1$ we are done. Similar analysis can be done for δ of type c. □

5.5.2. Proof for the d-hypergraph $\mathcal{G} = (\mathcal{V}, \mathcal{E})$ in the general case. We now reduce our hypergraph $\mathcal{G} = (\mathcal{V}, \mathcal{E})$ into two simpler hypergraphs $\mathcal{G}_+$ and $\mathcal{G}_{\mathrm{small}}$, as was done in the case of $\mathfrak{sl}_n$. Let $M = \{0, 1\}$ and set

$$\omega_2(e_\alpha) = \begin{cases} (d, 0) & \text{if } \alpha \in \Sigma^+(\mathfrak{g}, \mathfrak{t}) \\ (0, d^2) & \text{if } \alpha \in \Sigma^-(\mathfrak{g}, \mathfrak{t}) \end{cases} \text{ and } \omega_2(h_\alpha) = (0, d^2).$$

This coloring results in two sub-hypergraphs:

(1) $\mathcal{G}_+ = (\mathcal{V}_+, \mathcal{E}_+)$ consisting of vertices $\mathcal{V}_+ = \{e_\alpha\}_{\alpha \in \Sigma^+(\mathfrak{g}, \mathfrak{t})}$ and whose edges are

$$\mathcal{E}_+ = \{(\{e_{\alpha_{i_1}}, \ldots, e_{\alpha_{i_d}}\}, \widehat{e}_\delta) \in \mathcal{E} : \alpha_{i_k} \in \Sigma^+(\mathfrak{g}, \mathfrak{t}) \text{ for any } k \in [d]\}.$$

(2) $\mathcal{G}_{\mathrm{small}} = (\mathcal{V}_{\mathrm{small}}, \mathcal{E}_{\mathrm{small}})$ consisting of $\mathcal{E}_{\mathrm{small}} = \mathcal{E} \backslash \mathcal{E}_+$. By Proposition 5.26 and Lemma 5.28, it follows that all of the hyperedges in $\mathcal{E}_{\mathrm{small}}$ are of type $\widehat{e}_\delta$ with $\ell(\delta) < 4d$, and that $\mathcal{V}_{\mathrm{small}}$ consists of vertices $e_i \in \{h_\alpha\}_{\alpha \in \triangle} \cup \{e_\alpha\}_{\ell(\alpha) \leq 8d}$.

Proof for $\mathcal{G}_+$:

We apply level separation,

$$\omega_{\mathrm{LS},1}(e_\alpha) = \begin{cases} (2d,0) & \text{if } \left\lfloor \frac{\ell(\alpha)}{8} \right\rfloor \bmod 2 = 0 \\ (0,2d+1) & \text{if } \left\lfloor \frac{\ell(\alpha)}{8} \right\rfloor \bmod 2 = 1. \end{cases}$$

The difference from the coloring $\omega_{\mathrm{LS},1}$ in the case of $\mathfrak{sl}_n$ is that now the vertices $e_{\alpha_{i_1}}, \dots, e_{\alpha_{i_d}}$ of each hyperedge $(\{e_{\alpha_{i_1}}, \dots, e_{\alpha_{i_d}}\}, \widehat{e}_\delta)$ are not uniquely determined as in the $\mathfrak{sl}_n$ case (Lemma 5.28 versus Lemma 5.15). Note that each of $\mathrm{gr}_{\omega_{\mathrm{LS},1},0}\mathcal{G}_+$ and $\mathrm{gr}_{\omega_{\mathrm{LS},1},1}\mathcal{G}_+$ is a disjoint union of hypergraphs

$$\{\mathcal{G}_{+,u} = (\mathcal{V}_{+,u}, \mathcal{E}_{+,u})\}_{u\in\mathbb{N}},$$

where $\mathcal{E}_{+,u}$ consists of certain edges $\widehat{e}_\delta$ with $\ell(\delta) \in [L(u)+1, L(u)+12d]$, for some $L(u) \in \mathbb{N}$ depending on u.

Denote $\mathcal{G}_{+,u}$ by $\widetilde{\mathcal{G}}$ (as we deal with each of $\mathcal{G}_{+,u}$ in the same method). Using two level separations, we can separate these hypergraphs to ones that contain hyperedges $\widehat{e}_\delta$ with all δ of the same type (either a, b or c). Explicitly, set

$$\omega_{\mathrm{LS},2}(e_\alpha) = \begin{cases} (-2,-1,0) & \text{if } \alpha \text{ is of type } a \\ (0,-d,2-2d) & \text{if } \alpha \text{ is of type } b \\ (0,0,-3d) & \text{if } \alpha \text{ is of type } c, \end{cases}$$

and notice that by Lemma 5.25, $\omega_{\mathrm{LS},2}$ splits $\widetilde{\mathcal{G}}$ into three hypergraphs:

(1) $\widetilde{\mathcal{G}}_a := \mathrm{gr}_{\omega_{\mathrm{LS},2},1}(\widetilde{\mathcal{G}})$ which contains only edges $(\{e_{\alpha_{i_1}}, \dots, e_{\alpha_{i_d}}\}, \widehat{e}_\delta)$ with δ of type a.
(2) $\widetilde{\mathcal{G}}_b := \mathrm{gr}_{\omega_{\mathrm{LS},2},2}(\widetilde{\mathcal{G}})$ which contains only edges $(\{e_{\alpha_{i_1}}, \dots, e_{\alpha_{i_d}}\}, \widehat{e}_\delta)$ with δ of type b, and such that $\{\alpha_{i_1}, \dots, \alpha_{i_d}\}$ contains $d-1$ roots of type a and one root of type b.
(3) $\widetilde{\mathcal{G}}_{c,b} := \mathrm{gr}_{\omega_{\mathrm{LS},2},3}(\widetilde{\mathcal{G}})$ which contains only edges with δ of type c, and edges of type b such that $\{\alpha_{i_1}, \dots, \alpha_{i_d}\}$ contains $d-2$ roots of type a and two roots of type c (this only happens in the case of $\mathfrak{g} = \mathfrak{so}_{2n+1}$).

Using another coloring function (with two colors) on the hypergraph $\widetilde{\mathcal{G}}_{c,b}$ we can differentiate between the edges of type b and edges of type c, so that $\widetilde{\mathcal{G}}_{c,b}$ splits into $\widetilde{\mathcal{G}}'_b$ and $\widetilde{\mathcal{G}}_c$. Notice that $\widetilde{\mathcal{G}}_c$ is a hypergraph with at most $12d$ hyperedges (as there are at most $12d$ edges $\widehat{e}_\delta$ of type c with $\ell(\delta) \in [L(u)+1, L(u)+12d]$), so by Corollary 5.9, the graph $\widetilde{\mathcal{G}}_c$ is (FRS) after $12d^2$ convolutions. The graph $\widetilde{\mathcal{G}}_a$ can be solved as in the $\mathfrak{sl}_n$ case.

It is left to deal with $\widetilde{\mathcal{G}}_b$ and $\widetilde{\mathcal{G}}'_b$. We consider $\widetilde{\mathcal{G}}_b$ (the proof for $\widetilde{\mathcal{G}}'_b$ is similar). For any i let $\bar{i} = i \bmod 12d$ and for $s \in \{0, \dots, 12d-1\}$ set

$$\omega_{\mathrm{LS},3}(\epsilon_i \pm \epsilon_j)(s) := -i \cdot \delta_{s\bar{i}} \mp j \cdot \delta_{s\bar{j}},$$

and note that

$$\widetilde{\omega}_{\mathrm{LS},3}(\hat{e}_{b_{k,l}})(s) = -k \cdot \delta_{s\bar{k}} - l \cdot \delta_{s\bar{l}},$$

and in particular, the color of the hyperedge $\hat{e}_{b_{k,l}}$ is always $\bar{l}$ (with $k \le l$). As a consequence, for any $s \in \{0, \dots, 12d-1\}$ the hypergraph $\mathrm{gr}_{\omega_{\mathrm{LS},3},s}(\widetilde{\mathcal{G}}_b)$ consists only of hyperedges $\hat{e}_{b_{k,l}}$ with $l \bmod 12d = s$. Using a similar weight,

$$\omega_{\mathrm{LS},4}(\epsilon_i \pm \epsilon_j)(s) := n + (i-n)\delta_{s\bar{i}} \pm (j-n) \cdot \delta_{s\bar{j}},$$

we get

$$\widetilde{\omega}_{\mathrm{LS},4}(\hat{e}_{b_{k,l}})(s) = nd + (k-n)\cdot\delta_{s\bar{k}} + (l-n)\delta_{s\bar{l}},$$

and then for any $s_1, s_2 \in \{0,\dots,12d-1\}$ we get that

$$\widetilde{\mathcal{G}}_{b,s_1,s_2} := \mathrm{gr}_{\omega_{\mathrm{LS},4,s_2}}\left(\mathrm{gr}_{\omega_{\mathrm{LS},3,s_1}}(\widetilde{\mathcal{G}}_b)\right)$$

consists only of hyperedges $\hat{e}_{b_{k,l}}$ such that

$$l \bmod 12d = s_1 \text{ and } k \bmod 12d = s_2.$$

LEMMA 5.29. *Fix $s_1, s_2 \in \{0,\dots,12d-1\}$. Then all the hyperedges in $\widetilde{\mathcal{G}}_{b,s_1,s_2}$ are of types $b_{k,l}$ of the same length.*

PROOF. Note that $\ell(b_{k,l}) = 2n-l-k$ (resp. $2n-l-k+1$, $2n-l-k+2$) for $\mathfrak{g} = \mathfrak{so}_{2n}$ (resp. $\mathfrak{sp}_{2n}, \mathfrak{so}_{2n+1}$). But by the definition of $\mathcal{G}_{+,u}$, its edges $\mathcal{E}_{+,u}$ consists of $\widehat{e}_\delta$ with $\ell(\delta) \in [L(u)+1, L(u)+12d]$. Since $\hat{e}_{b_{k,l}} \in \widetilde{\mathcal{G}}_{b,s_1,s_2}$ it follows that $\ell(b_{k,l})$ has a unique value inside $[L(u)+1, L(u)+12d]$. □

Now for a fixed $s_1, s_2 \in \{0,\dots,12d-1\}$, let L be the unique length of the edges of $\widetilde{\mathcal{G}}_{b,s_1,s_2}$. Recall from Lemma 5.28, that the vertices of $\widetilde{\mathcal{G}}_{b,s_1,s_2}$ have lengths inside the interval $[\lfloor \frac{L}{d}\rfloor - 1, \lceil \frac{L}{d}\rceil + 1]$. Set $t := \lfloor \frac{L}{d}\rfloor - 2$. For any i let $\bar{i} = \lfloor \frac{i}{t}\rfloor \bmod 8d$. Set

$$\omega_{\mathrm{LS},5}(\epsilon_i \pm \epsilon_j)(s) := n + (i-n)\delta_{s\bar{i}} \pm (j-n)\cdot\delta_{s\bar{j}},$$

for $s \in \{0,\dots,8d-1\}$. Again we have

$$\widetilde{\omega}_{\mathrm{LS},5}(\hat{e}_{b_{k,l}})(s) = nd + (k-n)\cdot\delta_{s\bar{k}} + (l-n)\delta_{s\bar{l}},$$

and thus for any $s_3 \in \{0,\dots,8d-1\}$ we get that $\widetilde{\mathcal{G}}_{b,s_1,s_2,s_3} := \mathrm{gr}_{\omega_{\mathrm{LS},5,s_3}}(\widetilde{\mathcal{G}}_{b,s_1,s_2})$ consists only of hyperedges $\hat{e}_{b_{k,l}}$ such that

$$l \bmod 12d = s_1 \text{ and } k \bmod 12d = s_2 \text{ and } \left\lfloor \frac{k}{t} \right\rfloor \bmod 8d = s_3.$$

LEMMA 5.30. *For any $s_1, s_2 \in \{0,\dots,12d-1\}$ and $s_3 \in \{0,\dots,8d-1\}$, the hypergraph $\widetilde{\mathcal{G}}_{b,s_1,s_2,s_3}$ is a disjoint union of hypergraphs, each consisting of at most two hyperedges.*

PROOF. Let k_0 be minimal such that $\hat{e}_{b_{k_0,l_0}}$ is a hyperedge in $\widetilde{\mathcal{G}}_{b,s_1,s_2,s_3}$. Then the types of the hyperedges in $\widetilde{\mathcal{G}}_{b,s_1,s_2,s_3}$ are a subset of the types

$$\{b_{k_0+12jd,l_0-12jd}\}_{j\in\mathbb{N}},$$

where $k_0 + 12jd \leq l_0 - 12jd$ for any j. Furthermore, for any hyperedge $\hat{e}_{b_{k,l}}$ in $\widetilde{\mathcal{G}}_{b,s_1,s_2,s_3}$, k belongs to a specific interval $[t_0, t_0 + \lfloor \frac{L}{d}\rfloor - 1]$, which is roughly the length of a vertex of $\hat{e}_{b_{k,l}}$. Lemma 5.28 and the weight $\omega_{\mathrm{LS},5}$ guarantee that two edges $\hat{e}_{b_{k,l}}$ and $\hat{e}_{b_{k',l'}}$ share a common vertex only if $|k-k'| < 4d$, or $\left|k - l' - e\frac{L}{d}\right| < 4d$ for some $e \in \mathbb{Z}$. The assumption that $k = k' \bmod 12d$ and $\omega_{\mathrm{LS},5}$ guarantee that either $b_{k,l} = b_{k',l'}$ or there is precisely one $\hat{e}_{b_{k',l'}} \neq \hat{e}_{b_{k,l}}$ with a common vertex. □

Finally, using one more coloring function $\omega_{\mathrm{LS},6}$ (involving two colors) we may split $\widetilde{\mathcal{G}}_{b,s_1,s_2,s_3}$ to two hypergraphs, each a disjoint union of hyperedges. Moreover, each of these hyperedges is of type $b_{k,l}$ and by Lemma 5.28, the corresponding monomials (via Lemma 4.26) are of the form $x_1^{n_1} \cdot \dots \cdot x_k^{n_k}$ where

$$\sum_{i=1}^{k} n_i = d$$

and $n_i \leq 2$. By Corollary 5.9, $\dim V \geq r$ (resp. $\dim V \geq 3r$) is enough for each hyperedge to be flat (resp. (FRS)). In summary, we have

(1) The hypergraph $\widetilde{\mathcal{G}}_a$ is flat after $4d^3$ convolutions and (FRS) after $8d^3$ convolutions.
(2) The hypergraph $\widetilde{\mathcal{G}}_c$ is (FRS) (and flat) after $12d^2$ convolutions.
(3) The hypergraphs $\widetilde{\mathcal{G}'}_b$ and $\widetilde{\mathcal{G}}_b$ are flat after $12d \cdot 12d \cdot 8d \cdot 2 \cdot 1 = 2304d^3$ convolutions, and (FRS) after $12d \cdot 12d \cdot 8d \cdot 2 \cdot 3 = 6912d^3$ convolutions.

Altogether we get that $\mathcal{G}_+$ is flat after $10^4 d^3 > 2 \cdot \left(4d^3 + 12d^2 + 2 \cdot 2304d^3\right)$ convolutions and (FRS) after $3 \cdot 10^4 d^3$ convolutions.

Proof for $\mathcal{G}_{\text{small}}$:

Since in the general case we have a weaker version of Proposition 5.13 (namely Proposition 5.26), we need to apply a more careful analysis than in the $\mathfrak{sl}_n$ case. Set $\bar{i} = i \bmod 8d^2$ and $\omega_{\text{LS},7} : \mathcal{V}_{\text{small}} \to \mathbb{N}^{\{0,\dots,8d^2-1\}}$ by

$$\omega_{\text{LS},7}(\pm\epsilon_i \pm \epsilon_j)(s) := n \pm (i-n) \cdot \delta_{s\bar{i}} \pm (j-n) \cdot \delta_{s\bar{j}},$$

for $s \in \{0,1\}$. Note that both $\hat{e}_{a_{k,l}}, \hat{e}_{b_{k,l}}$ or $\hat{e}_{c_k}$ has the color $\bar{k}$ (with $k < l$). By Proposition 5.26 (and by the definition of $\mathcal{E}_{\text{small}}$), we have that $\operatorname{gr}_{\omega_{\text{LS},s}}(\mathcal{G}_{\text{small}})$ is a disjoint union of hypergraphs with at most $8d$ hyperedges. Using a similar coloring function with $8d$ colors, we reduce to a disjoint union of hyperedges (i.e., each hypergraph consists of hyperedges $\hat{e}_{a_{k,l}}, \hat{e}_{b_{k,l}}$ or $\hat{e}_{c_k}$ with $k = k_0 \bmod 8d^2$ and $l = l_0 \bmod 8d$ for some k_0, l_0). By Corollary 5.9, d convolutions are enough for the (FRS) property of a hyperedge.

Summary for $\mathcal{G}_{\text{small}}$ and for Γ_+ and Γ_-:

$\mathcal{G}_{\text{small}}$ is flat after $64d^3 = 8d^2 \cdot 8d \cdot 1$ convolutions and (FRS) after $64d^4 = 8d^2 \cdot 8d \cdot d$ convolutions. Combining with the case of $\mathcal{G}_+$ we get that:

(1) The hypergraphs Γ_+ and Γ_- are flat after $2 \cdot 10^4 d^3 > 10^4 d^3 + 64d^3$.
(2) The hypergraphs Γ_+ and Γ_- are (FRS) after $4 \cdot 10^4 d^4 > 3 \cdot 10^4 d^3 + 64d^4$.

5.5.3. Proof for Γ_0: The general case. The proof is almost identical to the case of $\mathfrak{sl}_n$ (see Subsection 5.3.3). Let $\mathfrak{g}$ be either $\mathfrak{so}_{2n}$, $\mathfrak{so}_{2n+1}$ or $\mathfrak{sp}_{2n}$. By eliminating edges containing vertices of type c, we may assume that $\mathfrak{g}$ is $\mathfrak{so}_{2n}$ or $\mathfrak{sp}_{2n}$. As in the case of $\mathfrak{sl}_n$, using an averaging weight, a monomialization weight, and the level separation method, we can reduce the m-th jet map $J_m(\Phi_{\Gamma_0,V,\{\tau_\gamma\}})$ to a disjoint union of morphisms to a $\dim\mathfrak{t} \cdot (2d+1)$-dimensional vector space.

Let $\mathfrak{sl}_n \hookrightarrow \mathfrak{g}$ be the map

$$A \mapsto \begin{pmatrix} A & 0 \\ 0 & -A^t \end{pmatrix}.$$

Note that $\triangle$ contains $n-1$ simple roots $\{a_{12}, \dots, a_{n-1,n}\}$ of type a, and one simple root ($b_n := b_{n,n}$ or $b_n := b_{n-1,n}$) of type b. Let $\mathfrak{h}$ be the simple Lie algebra generated by $e_{a_{n-d,n-d+1}}, \dots, e_{a_{n,n-1}}$ and e_{b_n} (of the same type as $\mathfrak{g}$). Using an elimination weight, we may eliminate all monomials involving variables outside of $\operatorname{span}\{\mathfrak{h}, \mathfrak{sl}_n\}$. From here, the proof is essentially the same as in the case of $\mathfrak{sl}_n$, where $\mathfrak{h}$ takes the role of $\text{span}\{e_{i,j}\}_{i,j\leq d}$. In particular, $(\Gamma_0, V, \{\tau_\gamma\})$ is flat if $\dim V \geq 12d^2 r$ and it is (FRS) for $\dim V \geq 60d^3 r$, as in the case of $\mathfrak{sl}_n$.

5.5.4. Proof of Theorems 4.9, 4.11 and 5.23. Theorem 5.23 follows from the analysis we did for Γ_+, Γ_- and Γ_0.

PROOF OF THEOREM 4.9. By Propositions 4.21 and 4.22 it is enough to prove the theorem in the case of self-convolutions of a single d-homogeneous word $w \in \mathcal{L}_r$ of pure type, at the price of twice the required number of convolutions. By Proposition 3.6 and Corollary 3.18, it is enough to prove we can obtain the flatness and (FRS) properties.

(1) For a simple Lie algebra $\mathfrak{g}$ with $\text{rk}(\mathfrak{g}) > 4d$, the map φ_w^{*t} is flat (resp. (FRS)) if $t \geq 4 \cdot 10^4 d^3$ (resp. $t \geq 8 \cdot 10^4 d^4$).
(2) For a simple Lie algebra $\mathfrak{g}$ with $\text{rk}(\mathfrak{g}) \leq 4d$, we have $\dim\mathfrak{g} \leq 100d^2$. By Corollary 5.8, the map φ_w^{*t} is (FRS) if $t \geq 100d^3$.

This finishes the proof. □

PROOF OF THEOREM 4.11. Let $w \in \mathcal{L}_r$ be a Lie algebra word of degree d and let $\mathfrak{g}$ be a simple K-Lie algebra. At first assume that $\text{rk}(\mathfrak{g}) < 50d$ (and thus $\dim(\mathfrak{g}) < 10^4 d^2$). Then for any $Y \in \mathfrak{g}$ the fiber $\varphi_w^{-1}(Y)$ is defined by polynomials of degree at most d. By Part (2) of Fact 2.8, we have $\text{lct}(\mathfrak{g}^r, \varphi_w^{-1}(Y)) \geq \frac{1}{d}$ and therefore by Lemma 3.23 $\varphi_w : \mathfrak{g}^r \to \mathfrak{g}$ is $\frac{1}{10^4 d^3}$-jet flat.

Now assume that $\text{rk}\mathfrak{g} \geq 50d$. Choose a Chevalley basis $\mathcal{B} = \{e_\beta\}_{\beta \in \Sigma(\mathfrak{g},\mathfrak{t})} \cup \{h_\alpha\}_{\alpha \in \mathfrak{g}}$ and let $\hat{\mathcal{B}}$ be its dual basis. Identify $\varphi_w : \mathfrak{g}^r \to \mathfrak{g}$ with $\psi_w : \mathfrak{g}^r \to \mathbb{A}^{\dim\mathfrak{g}}$ via $\mathcal{B}$ and $\hat{\mathcal{B}}$. Let $\pi_+ : \mathbb{A}^{\dim\mathfrak{g}} \to \mathbb{A}^N$ be the projection to the root subspaces $\{e_\beta\}$, such that β is a positive root and $\ell(\beta) > 4d$. By our assumption on $\mathfrak{g}$, we have $N > \frac{1}{4}\dim\mathfrak{g}$. Notice that by Subsection 5.5.2 $(\pi_+ \circ \psi_w)^{*t}$ is (FRS) for $t \geq 3 \cdot 10^4 d^3$. By Lemma 3.26, $\pi_+ \circ \psi_w$ is $\frac{1}{3 \cdot 10^4 d^3}$-jet flat, so for any $Y \in \mathbb{A}^{\dim\mathfrak{g}}$ we have by Fact 2.8,

$$\text{lct}(\mathfrak{g}^r, \psi_w^{-1}(Y)) \geq \text{lct}(\mathfrak{g}^r, (\pi_+ \circ \psi_w)^{-1}(\pi_+(Y))) \geq \frac{N}{3 \cdot 10^4 d^3} \geq \frac{\dim\mathfrak{g}}{2 \cdot 10^5 d^3}.$$

Altogether we get that w is $\frac{1}{2 \cdot 10^5 d^3}$-jet flat.

For matrix word map, the same analysis yields that $w \in \mathcal{A}_r$ is $\frac{1}{25d^3}$-jet flat. □

CHAPTER 6

The commutator map revisited

In this section we use the methods of Section 5.1.1 in order to provide a simple proof to the following theorem:

THEOREM 6.1. *Let $w_0 = xyx^{-1}y^{-1} \in F_r$ and $w = [X,Y] \in \mathcal{L}_r$ be the commutator words. Then for any semisimple algebraic K-group $\underline{G}$, with Lie algebra $\mathfrak{g}$, the corresponding word maps $\varphi_{w_0}(x,y) = xyx^{-1}y^{-1}$ and $\psi_w(X,Y) = [X,Y]$ satisfy*

(1) $\varphi_{w_0}^{*4} : \underline{G}^8 \to \underline{G}$ *is (FRS).*
(2) $\psi_w^{*4} : \mathfrak{g}^8 \to \mathfrak{g}$ *is (FRS).*

It was shown in [**AA16**], that $\varphi_{w_0}^{*21}$ is (FRS) for all semisimple algebraic groups $\underline{G}$, and that ψ_w^{*21} is (FRS) for all classical Lie algebras. By using a more efficient degenerations of graphs, these bounds were improved to 11 in [**Kap**]. It is conjectured that $\varphi_{w_0}^{*2}$ and ψ_w^{*2} are (FRS), and this conjecture was verified for SL_n in [**Bud21**]. Theorem 6.1 improves these bounds (outside of the SL_n and $\mathfrak{sl}_n$ cases), and its proof is significantly simpler. It also treats the case of exceptional Lie algebras, which was untreated before. Furthermore, in [**AA16**, **AA18**], the (FRS) property of $\varphi_{w_0}^{*21}$ was used to bound the representation growth of compact p-adic and arithmetic groups by a polynomial independent of the rank of the group. Theorem 6.1 improves these bounds (see Corollary 1.9).

By [**AA16**, Theorem IV], in order to prove Theorem 6.1, it is enough to show that $\psi_w^{*4} : \mathfrak{g}^4 \to \mathfrak{g}$ is (FRS) at $(0,\ldots,0)$. We start by proving the following theorem, which is interesting on its own.

THEOREM 6.2. *The map ψ_w^{*2} is flat with geometrically irreducible, reduced fibers.*

The flatness of $\varphi_{w_0}^{*2} : \underline{G}^4 \to \underline{G}$ follows from [**Li93**] and [**Sha09**, Theorem 2.8]. For Lie algebras, as far as we know, only flatness of $\mathfrak{sl}_n$ was previously known (see [**CB01**] and also [**Bud21**]).

PROOF. We may assume that $K = \mathbb{C}$. Let $\mathfrak{g}$ be a simple Lie algebra and $\langle \cdot, \cdot \rangle$ be the Killing form on $\mathfrak{g}$. We can endow $\mathfrak{g}$ with a $\mathbb{Z}$-structure, by choosing a Chevalley basis. For almost any prime p, the form $\langle \cdot, \cdot \rangle$ is non-degenerate on $\mathfrak{g}(\mathbb{F}_p)$ (see e.g. [**Sel67**]), and thus we have an identification $Z \mapsto \Psi_Z$ between $\mathfrak{g}(\mathbb{F}_p)$ and $\mathfrak{g}(\mathbb{F}_p)^\vee$, where $\Psi_Z(W) := \Psi(\langle Z, W \rangle)$ and Ψ is a fixed non-trivial additive character of $\mathbb{F}_p$. Set $\tau_w(X) := \frac{|\psi_w^{-1}(X)|}{|\mathfrak{g}(\mathbb{F}_p)|^2}$. In order to prove that ψ_w^{*2} is (FGI) it is enough to show

$$(\star) \qquad |\mathfrak{g}(\mathbb{F}_p)| \cdot \left| \tau_w^{*2}(0) - \frac{1}{|\mathfrak{g}(\mathbb{F}_p)|} \right| = O(p^{-1/2}),$$

for almost any prime p. Indeed, the Lang-Weil bounds and $(\star)$ imply that ψ_w^{*2} is (FGI) over 0. Note that τ_w is a symmetric measure (i.e. $\tau_w(X) = \tau_w(-X)$ for every

$X \in \mathfrak{g}(\mathbb{F}_p)$), so $\left|(\psi_w^{*2})^{-1}(0)\right| \geq \left|(\psi_w^{*2})^{-1}(X)\right|$ for every $X \in \mathfrak{g}(\mathbb{F}_p)$. This will imply that ψ_w^{*2} is (FGI), as otherwise, using Chebotarev's density theorem, we may find infinitely many primes p that contradict ($\star$) (see e.g. Theorem 8.4). Using Fourier transform $\mathcal{F}$, we may write,

$$\tau_w(X) = \frac{1}{|\mathfrak{g}(\mathbb{F}_p)|} \sum_{Z \in \mathfrak{g}(\mathbb{F}_p)} \mathcal{F}(\tau_w)(Z) \cdot \Psi_Z(X).$$

We then have,

$$\begin{aligned}
\mathcal{F}(\tau_w)(Z) &= \sum_{W \in \mathfrak{g}(\mathbb{F}_p)} \tau_w(W) \cdot \Psi_Z(W) \\
&= \frac{1}{|\mathfrak{g}(\mathbb{F}_p)|^2} \sum_{W \in \mathfrak{g}(\mathbb{F}_p)} \left|\psi_w^{-1}(W)\right| \cdot \Psi_Z(W) \\
&= \frac{1}{|\mathfrak{g}(\mathbb{F}_p)|^2} \sum_{W \in \mathfrak{g}(\mathbb{F}_p)} \sum_{\substack{X, Y \in \mathfrak{g}(\mathbb{F}_p) \\ [X,Y] = W}} \Psi_Z(W) \\
&= \frac{1}{|\mathfrak{g}(\mathbb{F}_p)|^2} \sum_{X, Y \in \mathfrak{g}(\mathbb{F}_p)} \Psi_Z([X,Y]) \\
&= \frac{1}{|\mathfrak{g}(\mathbb{F}_p)|^2} \sum_{X, Y \in \mathfrak{g}(\mathbb{F}_p)} \Psi(\langle Z, [X,Y]\rangle) \\
&= \frac{1}{|\mathfrak{g}(\mathbb{F}_p)|^2} \sum_{X, Y \in \mathfrak{g}(\mathbb{F}_p)} \Psi(\langle X, [Y,Z]\rangle) \\
&= \frac{\left|\mathrm{Cent}_{\mathfrak{g}(\mathbb{F}_p)}(Z)\right|}{|\mathfrak{g}(\mathbb{F}_p)|},
\end{aligned}$$

where the last equality follows from the Schur orthogonality relations (i.e. averaging any non-trivial character over $\mathfrak{g}(\mathbb{F}_p)$ yields 0). In particular, we have the following:

$$(\star\star) \qquad \tau_w^{*2}(0) = \frac{1}{|\mathfrak{g}(\mathbb{F}_p)|} \sum_{Z \in \mathfrak{g}(\mathbb{F}_p)} \left(\mathcal{F}(\tau_w)(Z)\right)^2 \cdot \Psi_Z(0) = \sum_{Z \in \mathfrak{g}(\mathbb{F}_p)} \frac{\left|\mathrm{Cent}_{\mathfrak{g}(\mathbb{F}_p)}(Z)\right|^2}{|\mathfrak{g}(\mathbb{F}_p)|^3}.$$

Notice that ($\star\star$) has a geometric description. It equals $\frac{|\Upsilon(\mathbb{F}_p)|}{|\mathfrak{g}(\mathbb{F}_p)|^3}$, where

$$\Upsilon := \{(X,Y,Z) \in \mathfrak{g}^3 : [X,Z] = [Y,Z] = 0\}.$$

Lemma 6.3. *For any* $\dim\mathfrak{g} \geq k \in \mathbb{N}$, *the set*

$$S_k := \{Z \in \mathfrak{g} : \dim\mathrm{Cent}_{\mathfrak{g}}(Z) = k\}$$

satisfies the following if $k < \dim\mathfrak{g}$,

$$\dim S_k \leq \dim\mathfrak{g} + \mathrm{rk}(\mathfrak{g}) - k,$$

and furthermore $\dim S_{\dim\mathfrak{g}} = 0$.

Proof. We have $\psi_w^{-1}(0) \supseteq \{(Z,W) \in S_k \times \mathfrak{g} : [Z,W] = 0\}$, so

$$\dim S_k \leq \dim\psi_w^{-1}(0) - k = \dim\mathfrak{g} + \mathrm{rk}(\mathfrak{g}) - k,$$

where the last equality is well known (e.g. [**Ric79**, Theorem A]). □

LEMMA 6.4. *For any $0 \neq Z \in \mathfrak{g}$, we have,*

$$\dim \mathrm{Cent}_{\mathfrak{g}}(Z) \leq \dim \mathfrak{g} - \mathrm{rk}(\mathfrak{g}) - 1.$$

PROOF. We use the fact that any proper subalgebra of a simple Lie algebra is of codimension at least $\mathrm{rk}(\mathfrak{g})$ where equality is achieved only for type A, in which the only subalgebra of codimension $\mathrm{rk}(\mathfrak{g})$ is a parabolic subalgebra (e.g. [**CX18**, p.47, Table 1], [**Stu91**]). For Z semisimple, $\mathrm{Cent}_{\mathfrak{g}}(Z)$ is a Levi subalgebra so it is properly contained in a parabolic subalgebra. For non-nilpotent elements, their centralizer is contained in the centralizer of their semisimple part. For nilpotent elements, their centralizer cannot contain a maximal torus, so we are done. □

PROPOSITION 6.5. *We have* $\dim \Upsilon = 2\dim \mathfrak{g}$ *and furthermore,*

$$|\Upsilon(\mathbb{F}_p)| = |\mathfrak{g}(\mathbb{F}_p)|^2 (1 + O(p^{-1/2})).$$

PROOF. Notice that

$$\Upsilon = \bigcup_{k=0}^{\dim \mathfrak{g}} \{(X, Y, Z) \in \Upsilon : Z \in S_k\}$$

so

$$\begin{aligned} \dim \Upsilon &= \max_k \dim\{(X, Y, Z) \in \Upsilon : Z \in S_k\} \\ &= \max_k (2k + \dim S_k) \\ &= \max(2\dim \mathfrak{g}, \max_{k < \dim \mathfrak{g}} (2k + \dim S_k)) \\ &\leq 2\dim \mathfrak{g} \end{aligned}$$

where the inequality follows by Lemmas 6.3 and 6.4. Furthermore, Lemma 6.4 also implies that $\dim\{(X, Y, Z) \in \Upsilon : Z \in S_k\} \leq 2\dim \mathfrak{g} - 1$ for $k \neq \dim \mathfrak{g}$. Thus $\mathfrak{g} \times \mathfrak{g} \times \{0\}$ is the only top dimensional irreducible component of Υ so the proposition follows. □

Proposition 6.5 implies that $(\star\star)$ can be written as

$$\sum_{Z \in \mathfrak{g}(\mathbb{F}_p)} \frac{\left|\mathrm{Cent}_{\mathfrak{g}(\mathbb{F}_p)}(Z)\right|^2}{|\mathfrak{g}(\mathbb{F}_p)|^3} = \frac{|\Upsilon(\mathbb{F}_p)|}{|\mathfrak{g}(\mathbb{F}_p)|^3} = \frac{1}{|\mathfrak{g}(\mathbb{F}_p)|}(1 + O(p^{-\frac{1}{2}})),$$

so $(\star)$ holds. This proves that ψ_w^{*2} is flat and has geometrically irreducible fibers.

Note that the fibers of ψ_w^{*2} are local complete intersections and in particular Cohen-Macaulay, so by the $R_0 + S_1$-criterion (e.g. [**Gro67**, Proposition 5.8.5]), in order to show that the fibers are reduced it is enough to show that they are generically reduced. Since the fibers are irreducible, it is enough to find a smooth point in $(\psi_w^{*2})^{-1}(Y)$ for every $Y \in \mathfrak{g}$. Note that ψ_w is generically smooth over $\mathfrak{g}$, i.e. smooth over an open (dense) set $U \subseteq \mathfrak{g}$. Since $U + U = \mathfrak{g}$, we may write $Y = Y_1 + Y_2$, with $Y_1, Y_2 \in U$. Take any points $(X_1, X_2) \in \psi_w^{-1}(Y_1)$ and $(X_3, X_4) \in \psi_w^{-1}(Y_2)$ and then ψ_w^{*2} is smooth at (X_1, X_2, X_3, X_4). This implies that (X_1, X_2, X_3, X_4) is a smooth point of $(\psi_w^{*2})^{-1}(Y)$ and we are done. We have finished proving Theorem 6.2. □

We now turn to the proof of Theorem 6.1.

Recall from Section 2.1 (and Example 2.4) that the m-th jet map $J_m(\psi_w) : J_m(\mathfrak{g}^2) \to J_m(\mathfrak{g})$ is given as follows (where a_i are some non-zero numbers)

$$\begin{aligned} &J_m(\psi_w)(X_1, X_2, X_1^{(1)}, X_2^{(1)}, \ldots, X_1^{(m)}, X_2^{(m)}) \\ &\quad = ([X_1, X_2], [X_1^{(1)}, X_2^{(0)}] + [X_1^{(0)}, X_2^{(1)}], \ldots, \sum_{i+j=m} a_i [X_1^{(i)}, X_2^{(j)}]). \end{aligned}$$

Using an averaging weight ω_{av} and a monomialization weight ω_{mon} as in Section 5.1.1, we reduce to a map of the following form:

$$\begin{aligned} &\widetilde{\phi}_m(X_1, X_2, X_1^{(1)}, X_2^{(1)}, \ldots, X_1^{(m)}, X_2^{(m)}) \\ &\quad := ([X_1, X_2], [X_1^{(1)}, X_2], [X_1^{(1)}, X_2^{(1)}], \ldots, [X_1^{(\lceil \frac{m}{2} \rceil)}, X_2^{(\lfloor \frac{m}{2} \rfloor)}]). \end{aligned}$$

Using the level separation method, similarly to Section 5.1, we set $\omega_{\mathrm{LS}}(X_1^{(u)}) = (10^u \cdot 2, 10^u \cdot 3)$ and $\omega_{\mathrm{LS}}(X_2^{(u)}) = (10^u \cdot 3, 10^u \cdot 1)$ such that $\mathrm{gr}_{\omega_{\mathrm{LS}},1}(\widetilde{\phi}_m)$ and $\mathrm{gr}_{\omega_{\mathrm{LS}},2}(\widetilde{\phi}_m)$ are maps of the form

$$\phi_m(Y_1, Y_2, \ldots, Y_{2m-1}, Y_{2m}) = ([Y_1, Y_2], [Y_3, Y_4], \ldots, [Y_{2m-1}, Y_{2m}]).$$

We therefore see that $J_m(\psi_w^{*2})$ can be degenerated to a morphism ϕ_m which consists of a tower of commutator maps with disjoint variables. By Theorem 6.2 it follows that ϕ_m^{*2} is flat, with geometrically irreducible, reduced fibers. By Proposition 2.10 this property is well behaved under deformations. Since ϕ_m^{*2} is a degeneration of $J_m(\psi_w^{*4})$, we deduce that $J_m(\psi_w^{*4})$ is flat and locally integral at $(0, \ldots, 0)$. Note that all irreducible components of $(J_m(\psi_w^{*4}))^{-1}(0)$ touch $(0, .., 0)$, so $(J_m(\psi_w^{*4}))^{-1}(0)$ is irreducible. By Corollary 3.12 this implies that ψ_w^{*4} is (FRS) as required.

CHAPTER 7

Local behavior of word maps and a lower bound on the log canonical threshold of their fibers

In this chapter we study certain singularity properties of word maps, in terms of the complexity class of w. The most natural and commonly used way to measure the complexity of a word w is its length $\ell(w)$ (e.g. $\ell(xyx^3y^{-1}) = 6$). An alternative way is to consider its degree $\deg(w)$, which we will soon define. We will see that in order to capture singularity properties of word maps, locally around $(e, \dots, e)$, it is best to consider the degree of word maps, while if we are interested in global behavior, it is better to consider the length of the word. Before we state the main theorems of this section, we define the degree and symbol of a word and of a word map.

7.1. The degree of a word and of a word map

DEFINITION 7.1. Let G be a group. The *lower central series* of G is the sequence

$$G = \Gamma_1(G) \supseteq \Gamma_2(G) \supseteq \cdots \supseteq \Gamma_k(G) \supseteq \dots$$

where $\Gamma_{k+1}(G) := [G, \Gamma_k(G)]$.

The associated graded abelian group

$$\mathrm{gr}(G) = \bigoplus_{k \geq 1} \mathrm{gr}_k(G),$$

where $\mathrm{gr}_k(G) = \Gamma_k(G)/\Gamma_{k+1}(G)$, has the structure of a graded Lie ring, where the Lie bracket is induced from the group commutator (e.g. [**MKS76**]). In particular, iterating this construction for $G = F_r$, we get an isomorphism $\mathrm{gr}(F_r) \simeq \mathcal{L}_r^{\mathbb{Z}}$, where $\mathcal{L}_r^{\mathbb{Z}}$ is the free Lie ring on r elements. We write $\mathrm{symb} : G \to \mathrm{gr}(G)$ for the natural symbol map.

DEFINITION 7.2. A word $w \in F_r$ is said to be of *degree* d if $\mathrm{symb}(w) \in \mathcal{L}_r^{\mathbb{Z}}$ is of degree d.

REMARK 7.3. For the n-th iterated commutator word

$$w_n(x, y) = [\dots[[x, y], y], \dots, y]$$

it is easy to see that $\ell(w_n)$ is exponential in the degree (which is $n + 1$). Set $\alpha(k) := \min\{\ell(w) : \deg(w) = k\}$. It was shown in [**MP10**, Theorem 1.2] that $\alpha(k) \geq k$, so $\ell(w) \geq \deg(w)$ for any $w \in F_r$. The asymptotic properties of $\alpha(k)$ have been studied in [**ET15**], where it was shown that $\alpha(k)$ grows slower then $k^{1.45}$.

7.1.1. The symbol of a word map. Let $\underline{G}$ be a simple algebraic K-group, $w \in F_r$ a word, and let $\varphi_w : \underline{G}^r \to \underline{G}$ be the corresponding word map. The tangent cone $C_e\underline{G}$ at $e \in G$ is the Lie algebra $\mathfrak{g}$. Using the linearization method (Corollary 2.13), we can choose minimal $l \in \mathbb{N}$ such that $D^l_{(e,\dots,e)}(\varphi_w)$ is non-trivial, and then the word map $\varphi_w : \underline{G}^r \to \underline{G}$ degenerates to a non-trivial map $D^l_{(e,\dots,e)}(\varphi_w) : \mathfrak{g}^r \to \mathfrak{g}$.

DEFINITION 7.4. For l minimal as above, we call $D^l_{(e,\dots,e)}(\varphi_w) : \mathfrak{g}^r \to \mathfrak{g}$ the *symbol* of the map $\varphi_w : \underline{G}^r \to \underline{G}$ and denote it by $\widetilde{\varphi}_w : \mathfrak{g}^r \to \mathfrak{g}$.

PROPOSITION 7.5. *The map $\widetilde{\varphi}_w : \mathfrak{g}^r \to \mathfrak{g}$ is a homogeneous Lie algebra word map of degree $l+1$.*

PROOF. φ_w is defined over a finitely generated field $K' \subset K$. Embed $K' \hookrightarrow \mathbb{C}$ and consider $\mathfrak{g}_{\mathbb{C}} := \mathfrak{g} \otimes_{K'} \mathbb{C}$ and the word map $\varphi_w : \underline{G}(\mathbb{C})^r \to \underline{G}(\mathbb{C})$. Recall that $\exp : \mathfrak{g}(\mathbb{C}) \to \underline{G}(\mathbb{C})$ is a local diffeomorphism, whose inverse is log. By a generalization of the Baker-Campbell-Hausdorff formula we have,

$$\log(\exp(X_1)\cdots\exp(X_m)) = \sum_{n=1}^{\infty} \frac{(-1)^{n-1}}{n} \sum_{\sum_j r_{ij}>0} \frac{[X_1^{r_{11}}\cdots X_m^{r_{1m}}\cdots X_1^{r_{n1}}\cdots X_m^{r_{nm}}]}{\sum_{i=1}^{n}(r_{i1}+\ldots+r_{im})\cdot\prod_{i,j}^{n,m} r_{ij}!},$$

where

$$[X_1^{r_{11}}\cdots X_m^{r_{1m}}\cdots X_1^{r_{n1}}\cdots X_m^{r_{nm}}] := [\underbrace{X_1,\cdots[X_1,}_{r_{11}\text{ times}}\cdots[\underbrace{X_m,\cdots[X_m,}_{r_{1m}\text{ times}}\cdots[\underbrace{X_1,\cdots[X_1,}_{r_{n1}\text{ times}}\cdots[\underbrace{X_m,\cdots,X_m}_{r_{nm}\text{ times}}]]\cdots].$$

Now, given $\varphi_w(g_1,\dots,g_r) = g_{i_1}^{\varepsilon_1}\cdot\ldots\cdot g_{i_m}^{\varepsilon_m}$ with $\varepsilon_j \in \{1,-1\}$, we have

$$\log\circ\varphi_w\circ(\exp,\dots,\exp)(X_1,\dots,X_r) = \log(\exp(\varepsilon_1 X_{i_1})\cdot\ldots\cdot\exp(\varepsilon_m X_{i_m}))$$

By the chain rule, and since $d_{(0,\dots,0)}\exp = \mathrm{Id}$ and $d_{(e,\dots,e)}\log = \mathrm{Id}$ we deduce that $D^l_{(e,\dots,e)}\varphi_w$ is a homogeneous Lie algebra word map of degree $l+1$. □

PROPOSITION 7.6. *Let $\varphi_w : \underline{G}^r \to \underline{G}$ be a word map. Then $\widetilde{\varphi}_w$ is a Lie algebra word map of degree $\leq \ell(w)$.*

PROOF. We may assume that $\underline{G}$ is simply connected and thus we have an embedding $\mathrm{SL}_2 \hookrightarrow \underline{G}$. By Proposition 7.5, it is enough to prove the claim for SL_2, which holds by a direct calculation. □

Note that we have two notions of symbol: $\mathrm{symb}(w)$ as the image of w in $\mathrm{gr}(F_r)$ (see Definition 7.2) and $\widetilde{\varphi}_w = D^l_{(e,\dots,e)}(\varphi_w)$, the symbol of the word map. If $\mathrm{rk}(\underline{G}) > \ell(w)$, then it turns out that these two notions agree, i.e. $\varphi_{\mathrm{symb}(w)} = \widetilde{\varphi}_w$.

PROPOSITION 7.7. *Let $w \in F_r$. Then for any simple algebraic group $\underline{G}$ with $\mathrm{rk}(\underline{G}) > \ell(w)$ its symbol $\widetilde{\varphi}_w : \mathfrak{g}^r \to \mathfrak{g}$ is identical to $\varphi_{\mathrm{symb}(w)}$. In particular, in this case w is of degree d if and only if φ_w has symbol of degree d.*

PROOF. Assume that w has degree d. By the identification $\mathrm{gr}(F_r) \simeq \mathcal{L}_r^{\mathbb{Z}}$ it follows that w can be written as $w = w_1 \cdot w_0$ with $w_0 \in \Gamma_{d+1}(F_r)$ and w_1 a product of iterated commutators of the form $[\ldots[[g_{i_1}, g_{i_2}], g_{i_3}], \dots, g_{i_d}]$. By Proposition 4.5, Corollary 4.8, and by the Baker-Campbell-Hausdorff formula, it follows that $\widetilde{\varphi}_{w_1} = \varphi_{\mathrm{symb}(w_1)}$. Since $\widetilde{\varphi}_{w_0}$ is of degree $> d$ it follows that

$$\widetilde{\varphi}_w = \widetilde{\varphi}_{w_1} = \varphi_{\mathrm{symb}(w_1)} = \varphi_{\mathrm{symb}(w)}.$$

□

7.2. Algebro-geometric results on word maps

We now state the main results of this section. In Theorem 7.8 we discuss local behavior of fibers of word maps at $(e,\dots,e)$ in terms of the degree of the words. In Theorem 7.9, we bound from below the log canonical threshold of the fibers of word maps $\varphi_w : \mathrm{SL}_n^r \to \mathrm{SL}_n$ in terms of $\ell(w)$.

THEOREM 7.8. *Let $d \in \mathbb{N}$, let $\underline{G}$ be a connected semisimple K-group, and let $\{\varphi_{w_i} : \underline{G}^{r_i} \to \underline{G}\}_{i=1}^m$ be a collection of non-trivial word maps with symbols $\widetilde{\varphi}_{w_i}$ of degree at most d. Then there exists $0 < C < 10^6$, such that for any $m \geq C \cdot d^4$ the morphism $\varphi_{w_1} * \dots * \varphi_{w_m}$ is (FRS) at $(e,\dots,e)$.*

PROOF. By Corollary 2.13 and Proposition 7.5, it is enough to prove the theorem for homogeneous Lie algebra word maps. Then the claim follows from Theorem 4.9. $\square$

THEOREM 7.9. *Let $w \in F_r$. Then $\varphi_w : \mathrm{SL}_n^r \to \mathrm{SL}_n$ is $\frac{1}{10^{17}\ell(w)^9}$-jet flat. In particular, $\mathrm{lct}(\mathrm{SL}_n^r, \varphi_w^{-1}(g)) \geq \frac{\dim \mathrm{SL}_n}{10^{17}\ell(w)^9}$ for every $g \in \mathrm{SL}_n$.*

We first prove Theorem 7.9 for SL_{2n}. Let $\varphi_w : \mathrm{SL}_{2n}^r \to \mathrm{SL}_{2n}$ be a word map. Recall that SL_{2n} contains two large commutative unipotent subgroups; the $n \times n$ right upper unipotent block

$$\mathrm{Mat}_n^+ := \left\{ \begin{pmatrix} I_n & A \\ 0 & I_n \end{pmatrix} : A \in \mathrm{Mat}_{n\times n} \right\},$$

and similarly

$$\mathrm{Mat}_n^- := \left\{ \begin{pmatrix} I_n & 0 \\ B & I_n \end{pmatrix} : B \in \mathrm{Mat}_{n\times n} \right\}.$$

Moreover, $\mathrm{Mat}_n^\pm$ boundedly generates SL_{2n}:

LEMMA 7.10. *Let $\varepsilon_i = (-1)^{i+1}$. The multiplication map $\psi : \prod_{i=1}^{13} \mathrm{Mat}_n^{\varepsilon_i} \to \mathrm{SL}_{2n}$ is a surjective morphism.*

PROOF. This follows from the proof of [**Kas07**, Lemma 2.1]. We show that any

$$g := \begin{pmatrix} A_{11} & A_{12} \\ A_{21} & A_{22} \end{pmatrix}$$

is in $\prod_{i=1}^{13} \mathrm{Mat}_n^{\varepsilon_i}$. Note that the row rank of

$$\begin{pmatrix} A_{12} \\ A_{22} \end{pmatrix}$$

is n. This means we can find $B_1 \in \mathrm{Mat}_n^+$ such that

$$B_1 g := \begin{pmatrix} A'_{11} & A'_{12} \\ A_{21} & A_{22} \end{pmatrix},$$

where $A'_{12} = A_{12} + (B_1)_{12}A_{22}$ is invertible. Taking

$$B_2 = \begin{pmatrix} I_n & 0 \\ (I_n - A_{22}) \cdot (A'_{12})^{-1} & I_n \end{pmatrix} \in \mathrm{Mat}_n^-$$

we get

$$B_2B_1g = \begin{pmatrix} A'_{11} & A'_{12} \\ A'_{21} & I_n \end{pmatrix}.$$

Now, we can find $B_3 \in \mathrm{Mat}_n^+$ and $B_4 \in \mathrm{Mat}_n^-$ such that

$$B_3B_2B_1gB_4 := \begin{pmatrix} A''_{11} & 0 \\ 0 & I_n \end{pmatrix}$$

where $A''_{11} \in \mathrm{SL}_n$. Since any element in $\mathrm{SL}_n(\overline{K})$ is a commutator, we can write $A''_{11} = [B, C]$ for $B, C \in \mathrm{SL}_n(\overline{K})$. By [**Kas07**, p.314] we see that

$$\begin{pmatrix} [B,C] & 0 \\ 0 & I_n \end{pmatrix} \in \psi(\prod_{i=1}^{9} \mathrm{Mat}_n^{\varepsilon_i}),$$

so we are done. □

Denote $\Phi_w := \varphi_w \circ (\psi * \psi, \dots, \psi * \psi) : (\prod_{i=1}^{13} \mathrm{Mat}_n^{\varepsilon_i} \times \prod_{i=1}^{13} \mathrm{Mat}_n^{\varepsilon_i})^r \to \mathrm{SL}_{2n}$.

LEMMA 7.11. *For any $\varepsilon > 0$, if Φ_w is ε-jet flat, then φ_w is ε-jet flat.*

PROOF. Assume that Φ_w is ε-jet flat. It is enough to show that $J_m(\varphi_w)$ is ε-flat over $\mathrm{SL}_{2n} \subset J_m(\mathrm{SL}_{2n})$. Since ψ is surjective, $J_m(\psi)$ is dominant, thus $J_m(\psi * \psi)$ is surjective for each $m \in \mathbb{N}$. We get that for every $g \in \mathrm{SL}_{2n}$,

$$\begin{aligned}\dim J_m(\mathrm{SL}_{2n}^r) - \dim J_m\left((\varphi_w)^{-1}(g)\right) &\geq \dim J_m(\prod_{i=1}^{13} \mathrm{Mat}_n^{\varepsilon_i})^{2r} - \dim J_m\left((\Phi_w)^{-1}(g)\right) \\ &\geq \varepsilon \dim J_m(\mathrm{SL}_{2n})\end{aligned}$$

so φ_w is ε-jet flat. □

PROPOSITION 7.12. *The map Φ_w is of the form*

$$\Phi_w(A_1, \dots, A_{26r}) = \begin{pmatrix} P_{11}(A_1, \dots, A_{26r}) & P_{12}(A_1, \dots, A_{26r}) \\ P_{21}(A_1, \dots, A_{26r}) & P_{22}(A_1, \dots, A_{26r}) \end{pmatrix},$$

where P_{ij} are matrix word maps of degree at most $26\ell(w)$.

PROOF. Notice that any word map $\varphi_w : \underline{G}^r \to \underline{G}$ is of the form $\varphi_w(g_1, \dots, g_r) = g_{i_1}^{\varepsilon_1} \dots g_{i_{\ell(w)}}^{\varepsilon_{\ell(w)}}$, with $\varepsilon_{i_k} \in \{\pm 1\}$, since the inverse of

$$\begin{pmatrix} I & A \\ 0 & I \end{pmatrix} \text{ is } \begin{pmatrix} I & -A \\ 0 & I \end{pmatrix}$$

we get that Φ_w has the above shape. □

COROLLARY 7.13. *The map $\varphi_w : \mathrm{SL}_{2n}^r \to \mathrm{SL}_{2n}$ is $\frac{1}{2\cdot 10^6 \ell(w)^3}$-jet flat for any $w \in F_r$.*

PROOF. $\Phi_w^{-1}(e) \subseteq P_{12}(A_1, \dots, A_{26r})^{-1}(0)$ so by Item (1) of Fact 2.8 we get the following:

$$\mathrm{lct}((\prod_{i=1}^{13} \mathrm{Mat}_n^{\varepsilon_i})^{2r}, \Phi_w^{-1}(e)) \geq \mathrm{lct}((\prod_{i=1}^{13} \mathrm{Mat}_n^{\varepsilon_i})^{2r}, P_{12}^{-1}(0)).$$

By Proposition 7.12, P_{12} is a matrix word map of degree at most $26\ell(w)$ so by Theorem 4.11

$$\operatorname{lct}((\prod_{i=1}^{13}\mathrm{Mat}_n^{\varepsilon_i})^{2r}, P_{12}^{-1}(0)) \geq \frac{\dim\mathrm{Mat}_n}{25\,(26\ell(w))^3} \geq \frac{\dim\mathrm{SL}_{2n}}{2\cdot 10^6\ell(w)^3}.$$

Thus, Φ_w is $\frac{1}{2\cdot 10^6\ell(w)^3}$-jet flat, and by Lemma 7.11 φ_w is $\frac{1}{2\cdot 10^6\ell(w)^3}$-jet flat. □

Proof of Theorem 7.9. We may assume that $r < \ell(w)$. We have proved the theorem for SL_{2n}. Consider $\varphi_{w,2n-1} : \mathrm{SL}_{2n-1}^r \to \mathrm{SL}_{2n-1}$ and note that $\varphi_{w,2n-1}^{-1}(g) \subseteq \varphi_{w,2n}^{-1}(g)$ for any $g \in \mathrm{SL}_{2n-1}$. In particular, by Corollary 7.13, if $n \geq 10^7\ell(w)^3 r$ then

$$\begin{aligned}
\sup_{m\geq 0}\frac{\dim J_m\left(\varphi_{w,2n-1}^{-1}(g)\right)}{m+1} &\leq \sup_{m\geq 0}\frac{\dim J_m\left(\varphi_{w,2n}^{-1}(g)\right)}{m+1}\\
&\leq \dim\mathrm{SL}_{2n}\left(r - \frac{1}{2\cdot 10^6\ell(w)^3}\right)\\
&\leq \dim\mathrm{SL}_{2n-1}\left(r - \frac{1}{2\cdot 10^6\ell(w)^3}\right) + 4nr\\
&\leq \dim\mathrm{SL}_{2n-1}\left(r - \frac{1}{10^6\ell(w)^3}\right).
\end{aligned}$$

Now assume that $n < 10^7\ell(w)^3 r < 10^7\ell(w)^4$. By [**Kas07**, Lemma 2.1] (and as explained in Lemma 7.13), any $g \in \mathrm{SL}_{2n-1}$ can be written as a product of at most 20 matrices of the form

$$\begin{pmatrix} I_p & 0\\ B & I_q\end{pmatrix} \text{ or } \begin{pmatrix} I_p & A\\ 0 & I_q\end{pmatrix},$$

where $A \in \mathrm{Mat}_{p\times q}$ and $B \in \mathrm{Mat}_{q\times p}$. Similarly as in the case of SL_{2n}, and by Lemma 7.11, this yields a map $\Phi_w : \mathbb{A}^N \to \mathrm{SL}_{2n-1}$, such that

$$\operatorname{lct}(\mathrm{SL}_{2n-1}^r, \varphi_{w,2n-1}^{-1}(g)) \geq \operatorname{lct}(\mathbb{A}^N, \Phi_w^{-1}(g)).$$

Since Φ_w is a polynomial map of degree at most $40\ell(w)$, we have by Item (2) of Fact 2.8,

$$\operatorname{lct}(\mathrm{SL}_{2n-1}^r, \varphi_{w,2n-1}^{-1}(g)) \geq \frac{1}{40\ell(w)} \geq \frac{\dim\mathrm{SL}_{2n-1}}{10^{17}\ell(w)^9}.$$ □

CHAPTER 8

Number theoretic interpretation of the flatness, ε-jet flatness and (FRS) properties

Let $\varphi : X \to Y$ be a dominant morphism between smooth, geometrically irreducible $\mathbb{Q}$-varieties. In this section, we provide number theoretic interpretations to the flatness, (FRS) and ε-jet flatness properties of φ. Flatness (and ε-flatness) can be interpreted, via the Lang-Weil bounds, as certain uniform bounds on the fibers of $\varphi : X(\mathbb{F}_q) \to Y(\mathbb{F}_q)$ over all finite fields. The (FRS) and ε-jet flatness properties can be interpreted by certain uniform bounds on the fibers over finite rings of the form $\mathbb{Z}/p^k\mathbb{Z}$. For the ε-jet flatness property, the idea is to apply the Lang-Weil bounds to the k-th jet maps $J_k(\varphi)$, and to use results in motivic integration to deal with the fact that the family of morphisms $\{J_k(\varphi)\}_{k\in\mathbb{N}}$ is of unbounded complexity.

8.1. The Lang-Weil bounds and a number theoretic characterization of flatness

DEFINITION 8.1 (cf. [**GH19**, Definition 7.7]). Let K' be any field.

(1) An affine K'-scheme X has *complexity at most* M, if X has a closed embedding $\psi_X : X \hookrightarrow \mathbb{A}^n_{K'}$ with $X = \operatorname{Spec}(K'[x_1, \dots, x_n]/(f_1, \dots, f_k))$, where n, k and $\max\limits_i\{\deg(f_i)\}$ are at most M.

(2) A morphism $\varphi : X \to Y$ of affine K'-schemes is said to be *of complexity at most* M, if there exist embeddings $\psi_X : X \hookrightarrow \mathbb{A}^{n_1}_{K'}$ and $\psi_Y : Y \hookrightarrow \mathbb{A}^{n_2}_{K'}$ of complexity at most M, such that the polynomial map $\widetilde{\varphi} : \mathbb{A}^{n_1}_{K'} \to \mathbb{A}^{n_2}_{K'}$ induced from φ is of the form $\widetilde{\varphi} = (\widetilde{\varphi}_1, \dots, \widetilde{\varphi}_{n_2})$, where the degrees of $\{\widetilde{\varphi}_i\}_{i=1}^{n_2}$ are at most M.

This definition can be extended to non-affine schemes, and morphisms between them (see e.g. [**GH19**, Definition 7.7]). We denote by $\mathcal{C}_M$ the class of all K'-schemes and K'-morphisms of complexity at most M. Since K' is usually clear from the context, we omit it from the notation.

DEFINITION 8.2. Let Z be a finite type $\mathbb{F}_q$-scheme, and let $\varphi : X \to Y$ be a morphism between finite type $\mathbb{Z}$-schemes.

(1) We denote by C_Z the number of top-dimensional geometrically irreducible components of Z which are defined over $\mathbb{F}_q$.

(2) We write $C_{X,q}{:=}C_{X_{\mathbb{F}_q}}$ and $C_{\varphi,q,y} := C_{(X_{\mathbb{F}_q})_{y,\varphi}}$ for any $y \in Y(\mathbb{F}_q)$.

THEOREM 8.3 (The Lang-Weil estimates [**LW54**]). *For any $M \in \mathbb{N}$, there exists $C(M) > 0$, such that for each finite type $\mathbb{F}_q$-scheme $X \in \mathcal{C}_M$ we have*

$$\left| \frac{|X(\mathbb{F}_q)|}{q^{\dim X}} - C_X \right| < C(M) q^{-\frac{1}{2}}.$$

For a set S of primes, we set $\mathcal{P}_S := \{q \in \mathbb{N} : q = p^r \text{ for a prime } p \notin S \text{ and } r \in \mathbb{N}\}$. The following is a relative version of the Lang-Weil bounds which will be very useful later on.

THEOREM 8.4. *Let $\varphi : X \to Y$ be a dominant morphism between finite type $\mathbb{Z}$-schemes X and Y such that $X_{\mathbb{Q}}$ and $Y_{\mathbb{Q}}$ are smooth and geometrically irreducible. Then the following are equivalent:*

(1) *There exist $C_1 > 0$ and a set S primes, such that for every $p \notin S$, and every $y \in Y(\mathbb{F}_p)$*
$$\left|\frac{|\varphi^{-1}(y)|}{p^{(\dim X_{\mathbb{Q}} - \dim Y_{\mathbb{Q}})}} - C_{\varphi,p,y}\right| < C_1 p^{-\frac{1}{2}}.$$
(2) *There exist $C_2 > 0$ and a set S of primes, such that for every $q \in \mathcal{P}_S$, and every $y \in Y(\mathbb{F}_q)$*
$$\left|\frac{|\varphi^{-1}(y)|}{q^{(\dim X_{\mathbb{Q}} - \dim Y_{\mathbb{Q}})}} - C_{\varphi,q,y}\right| < C_2 q^{-\frac{1}{2}}.$$
(3) *There exist $C_3 > 0$ and a set S of primes, such that for every $q \in \mathcal{P}_S$, it holds that*
$$\left|\frac{\varphi_*(\mu_{X(\mathbb{F}_q)})}{\mu_{Y(\mathbb{F}_q)}} - \widetilde{C}_{\varphi,q}\right|_{L^\infty} < C_3 q^{-\frac{1}{2}},$$
where $\widetilde{C}_{\varphi,q} : Y(\mathbb{F}_q) \to \mathbb{N}$ is defined by $\widetilde{C}_{\varphi,q}(y) := C_{\varphi,q,y}$.
(4) *$\varphi_{\mathbb{Q}} : X_{\mathbb{Q}} \to Y_{\mathbb{Q}}$ is flat.*

PROOF. We would like to show (1) $\Rightarrow$ (4) $\Rightarrow$ (2) $\Rightarrow$ (1) and (3) $\Leftrightarrow$ (2). Clearly (2) $\Rightarrow$ (1).

(3) $\Leftrightarrow$ (2). We prove (2) $\Rightarrow$ (3), the proof of (3) $\Rightarrow$ (2) is similar. Choose S large enough such that $\dim X_{\mathbb{Q}} = \dim X_{\mathbb{F}_q}$, $\dim Y_{\mathbb{Q}} = \dim Y_{\mathbb{F}_q}$ and such that $X_{\mathbb{F}_q}$ and $Y_{\mathbb{F}_q}$ are geometrically irreducible for every $q \in \mathcal{P}_S$. By the triangle inequality it is enough to show that
$$\left|\frac{|\varphi^{-1}(y)|}{q^{(\dim X_{\mathbb{Q}} - \dim Y_{\mathbb{Q}})}} - \frac{\varphi_*(\mu_{X(\mathbb{F}_q)})}{\mu_{Y(\mathbb{F}_q)}}(y)\right| < C_4 q^{-\frac{1}{2}}, \tag{8.1}$$
for any $y \in Y(\mathbb{F}_q)$. The left hand side of (8.1) can be written as
$$\begin{aligned}|\varphi^{-1}(y)|\left|\frac{1}{q^{(\dim X_{\mathbb{Q}} - \dim Y_{\mathbb{Q}})}} - \frac{|Y(\mathbb{F}_q)|}{|X(\mathbb{F}_q)|}\right| &< \frac{C_5\left|\varphi^{-1}(y)\right|}{q^{(\dim X_{\mathbb{Q}} - \dim Y_{\mathbb{Q}})}} q^{-\frac{1}{2}} \\ &< C_5 q^{-\frac{1}{2}}(C_{\varphi,q,y} + C_2 q^{-\frac{1}{2}}) \\ &< C_4 q^{-\frac{1}{2}},\end{aligned}$$
where the first inequality follows from our assumption on S and the Lang-Weil bounds (Theorem 8.3), and the last inequality follows from the fact that the collection $\{C_{\varphi,q,y}\}_{y,q}$ is uniformly bounded (see e.g. [**Gro67**, Corollary 9.7.9]).

(4) $\Rightarrow$ (2). Assume that $\varphi_{\mathbb{Q}}$ is flat. Then there exists a set S of primes such that for any $q \in \mathcal{P}_S$, $\varphi_{\mathbb{F}_q}$ is flat, $\dim X_{\mathbb{Q}} = \dim X_{\mathbb{F}_q}$ and $\dim Y_{\mathbb{Q}} = \dim Y_{\mathbb{F}_q}$. Since all of the fibers of $\varphi_{\mathbb{F}_q}$ belong to $\mathcal{C}_M$, for some $M \in \mathbb{N}$, the Lang-Weil estimates (Theorem 8.3) imply (2).

(1) $\Rightarrow$ (4). Since $X_{\mathbb{Q}}$ and $Y_{\mathbb{Q}}$ are smooth, by enlarging S, we may assume that $X_{\mathbb{F}_p}$ and $Y_{\mathbb{F}_p}$ are smooth for every $p \notin S$. Using miracle flatness it is enough to show that $\dim X_{\varphi(x)} = \dim X_{\mathbb{Q}} - \dim Y_{\mathbb{Q}}$ for every $x \in X(\overline{\mathbb{Q}})$.

Given $x \in X(\overline{\mathbb{Q}})$, there exists a finite extension $K/\mathbb{Q}$ such that $x \in X(K)$ and such that the geometrically irreducible components of the fiber $X_{\varphi(x)}$ are defined over K. Note that $\varphi(x) \in Y(S_1^{-1}\mathcal{O}_K)$, where $\mathcal{O}_K$ is the ring of integers of K, and $S_1 \subset \operatorname{Spec}(\mathcal{O}_K)$ is a finite set of primes. Write $Z = (X_{S_1^{-1}\mathcal{O}_K})_{\varphi(x)}$. Since $\operatorname{Spec}(K)$ is the generic point of $\operatorname{Spec}(S_1^{-1}\mathcal{O}_K)$, by [**Gro67**, Proposition 9.7.8] (see also [**AA18**, Theorem 2.3.5]), for all but finitely many primes $[\mathfrak{p}] \in \operatorname{Spec}(S_1^{-1}\mathcal{O}_K)$ it holds that all irreducible components of $Z_{[\mathfrak{p}]}$ are geometrically irreducible.

Further note that $Z_{[\mathfrak{p}]} \simeq X_{\varphi(\overline{x})}$ where $\overline{x}$ is the image of x under the natural map $X(S_1^{-1}\mathcal{O}_K) \to X(S_1^{-1}\mathcal{O}_K/\mathfrak{p})$. Now, using Chebotarev's density theorem, there exist infinitely many primes $[\mathfrak{p}] \in \operatorname{Spec}(S_1^{-1}\mathcal{O}_K)$ such that $S_1^{-1}\mathcal{O}_K/\mathfrak{p} \simeq \mathbb{F}_p$, for (infinitely many) prime numbers p. Since by our construction $C_{\varphi,p,\varphi(\overline{x})} \neq 0$, using our assumption and the Lang-Weil estimates we conclude $\dim Z_{[\mathfrak{p}]} = \dim X_{\varphi(\overline{x})} = \dim X_{\mathbb{Q}} - \dim Y_{\mathbb{Q}}$ for infinitely many primes $[\mathfrak{p}] \in \operatorname{Spec}(S_1^{-1}\mathcal{O}_K)$. This implies $\dim X_{\varphi(x)} = \dim X_{\mathbb{Q}} - \dim Y_{\mathbb{Q}}$. $\square$

8.2. Number theoretic and analytic interpretation of the (FRS) property

In [**AA16**], Aizenbud and Avni gave an analytic criterion for the (FRS) property:

THEOREM 8.5 (cf. [**AA16**, Theorem 3.4]). *Let $\varphi : X \to Y$ be a map between smooth $\mathbb{Q}$-varieties. Then the following are equivalent:*

(1) *φ is (FRS).*
(2) *For any non-Archimedean local field $\mathbb{Q} \subseteq F$ and any smooth, compactly supported measure μ on $X(F)$, the measure $\varphi_*(\mu)$ has continuous density.*
(3) *For any $x \in X(\overline{\mathbb{Q}})$ and any finite extension $K/\mathbb{Q}$ with $x \in X(K)$, there exists a non-Archimedean local field $F \supseteq K$ and a non-negative smooth, compactly supported measure μ on $X(F)$ that does not vanish at x such that $\varphi_*(\mu)$ has continuous density.*

In [**AA18**], a number theoretic characterization was given for the property of having rational singularities in the case of local complete intersection schemes. A strengthening of this characterization was later given in [**Gla19**].

DEFINITION 8.6. Let X be a finite type $\mathbb{Z}$-scheme. For a finite ring A, set $h_X(A) := \frac{|X(A)|}{|A|^{\dim X_{\mathbb{Q}}}}$ if $X_{\mathbb{Q}}$ is non-empty, and $h_X(A) = 0$ otherwise.

THEOREM 8.7 (see [**AA18**, Theorem A] and [**Gla19**, Theorem 1.4]). *Let X be a finite type $\mathbb{Z}$-scheme such that $X_{\mathbb{Q}}$ is equidimensional and a local complete intersection. Then the following are equivalent:*

(1) *For any k, $\lim_{p \to \infty} h_X(\mathbb{Z}/p^k\mathbb{Z}) = 1$.*
(2) *$X_{\overline{\mathbb{Q}}}$ is irreducible, and there exists a finite set S of primes, such that for every prime $p \notin S$ the sequence $k \mapsto h_X(\mathbb{Z}/p^k\mathbb{Z})$ is bounded.*
(3) *There exists $C > 0$ such that $\left|h_X(\mathbb{Z}/p^k\mathbb{Z}) - 1\right| < Cp^{-1/2}$ for every prime p and every $k \in \mathbb{N}$.*

(4) $X_{\overline{\mathbb{Q}}}$ *is reduced, irreducible and has rational singularities.*

REMARK 8.8. Notice that an (FRS) morphism $\varphi : X \to Y$ between geometrically irreducible, smooth $\mathbb{Q}$-varieties is in particular flat, so its fibers are local complete intersections. The condition of having rational singularities puts each individual $\mathbb{Z}$-fiber of φ in the framework of Theorem 8.7. Two difficulties now arise when trying to provide a version of Theorem 8.7 for families. Firstly, it is not enough to consider $\mathbb{Z}$-fibers (e.g., we might have $Y(\mathbb{Z}) = \varnothing$). Secondly, Theorem 8.7 gives us a different bound $C'(y)$ for each fiber X_y where $y \in Y(\mathbb{Z})$, and it is a priori not clear why it should be uniform. These difficulties are dealt with in [**CGH**], where a uniform bound is given on the size of fibers of (FRS) morphisms (see [**CGH**, Theorem A]).

8.3. The Denef-Pas language and motivic functions

The *Presburger language*, denoted

$$\mathcal{L}_{\text{Pres}} = (+, -, \leq, \{\equiv_{\text{mod } n}\}_{n>0}, 0, 1)$$

is a first order language of ordered abelian groups along with constants $0, 1$ and a family of relations $\{\equiv_{\text{mod } n}\}_{n>0}$ of congruences modulo n. The *Denef-Pas language*, denoted

$$\mathcal{L}_{\text{DP}} = (\mathcal{L}_{\text{Val}}, \mathcal{L}_{\text{Res}}, \mathcal{L}_{\text{Pres}}, \text{val}, \text{ac})$$

is a first order three sorted language, where

(1) The valued field sort VF is endowed with the language of rings $\mathcal{L}_{\text{Val}} = (+, -, \cdot, 0, 1)$.
(2) The residue field sort RF is endowed with the language of rings $\mathcal{L}_{\text{Res}} = (+, -, \cdot, 0, 1)$.
(3) The value group sort VG (which we just call $\mathbb{Z}$), is endowed with the Presburger language $\mathcal{L}_{\text{Pres}} = (+, -, \leq, \{\equiv_{\text{mod } n}\}_{n>0}, 0, 1)$.
(4) $\text{val} : \text{VF}\backslash\{0\} \to \mathbb{Z}$ and $\text{ac} : \text{VF} \to \text{RF}$ are two function symbols.

Let Loc be the collection of all non-Archimedean local fields. For $F \in \text{Loc}$, we denote by k_F its residue field, and set $q_F := |k_F|$. We use the notation $\text{Loc}_{\gg}$ for the collection of $F \in \text{Loc}$ with large enough residual characteristic. By choosing a uniformizer π of $\mathcal{O}_F$, we can interpret val and ac as the usual valuation map $\text{val} : F^{\times} \to \mathbb{Z}$ and angular component map $\text{ac} : F \to k_F$. Hence, we can interpret any formula ϕ in $\mathcal{L}_{\text{DP}}$ with n_1 free VF-variables, n_2 free RF-variables and n_3 free $\mathbb{Z}$-variables, in $F \in \text{Loc}$, yielding a subset $\phi(F) \subseteq F^{n_1} \times k_F^{n_2} \times \mathbb{Z}^{n_3}$. A collection $X = (X_F)_{F \in \text{Loc}_{\gg}}$ with $X_F = \phi(F)$ is called an $\mathcal{L}_{\text{DP}}$*-definable set.* Given definable sets X and Y, an $\mathcal{L}_{\text{DP}}$*-definable function* is a collection $f = (f_F : X_F \to Y_F)_{F \in \text{Loc}}$ of functions whose collection of graphs $(\Gamma_{f_F})_{F \in \text{Loc}}$ is a definable set.

The class of definable functions is not preserved under integration, which motivates us to consider a larger class of functions.

DEFINITION 8.9 (See [**CGH16**, definitions 2.3-2.6]). Let X be a definable set. A collection $h = (h_F)_{F \in \text{Loc}}$ of functions $h_F : X_F \to \mathbb{R}$ is called a *motivic* function, if for $F \in \text{Loc}_{\gg}$ it can be written as:

$$h_F(x) = \sum_{i=1}^{N_1} |Y_{i,F,x}| q_F^{\alpha_{i,F}(x)} \left(\prod_{j=1}^{N_2} \beta_{ij,F}(x) \right) \left(\prod_{j=1}^{N_3} \frac{1}{1 - q_F^{a_{ij}}} \right),$$

where:

- N_1, N_2, N_3 and a_{il} are non-zero integers.
- $\alpha_i : X \to \mathbb{Z}$ and $\beta_{ij} : X \to \mathbb{Z}$ are definable functions.
- $Y_{i,F,x} = \{\xi \in k_F^{r_i} : (x,\xi) \in Y_{i,F}\}$ is the fiber over x of a definable set $Y_i \subseteq X \times \mathrm{RF}^{r_i}$ with $r_i \in \mathbb{N}$.

We denote by $\mathcal{C}(X)$ the ring of motivic functions on a definable set X. Unlike the class of definable functions, this class is preserved under integration:

THEOREM 8.10 ([**CGH14**, Theorem 4.3.1]). *Let X be an $\mathcal{L}_{\mathrm{DP}}$-definable set, and let $f \in \mathcal{C}(X \times \mathrm{VF}^m)$. Then there exists a function $g \in \mathcal{C}(X)$ such that for every $F \in \mathrm{Loc}_{\gg}$ and $x \in X_F$, if $f_F(x,y) \in L^1(F^m)$ then*

$$g_F(x) = \int_{y \in F^m} f_F(x,y)dy.$$

REMARK 8.11. One can also define motivic measures and functions on algebraic varieties, and show that this class is preserved under integration. For more details, see for example [**GH19**, Section 3.3].

We further need the following structure theory for $\mathcal{L}_{\mathrm{Pres}}$-definable functions.

DEFINITION 8.12 (See [**Clu03**, Definition 1] and [**CL08**, Section 4.1]). Let $X \subseteq \mathbb{Z}^m$ be an $\mathcal{L}_{\mathrm{Pres}}$-definable set. An $\mathcal{L}_{\mathrm{Pres}}$-definable function $f : X \to \mathbb{Z}$ is *linear* if there is $\gamma \in \mathbb{Z}$ and integers a_i and $0 \leq c_i < n_i$ for $1 \leq i \leq m$ such that $x_i - c_i \equiv 0 \bmod n_i$ and

$$f(x_1, \ldots, x_m) = \sum_{i=1}^{m} a_i(\frac{x_i - c_i}{n_i}) + \gamma.$$

It turns out that every $\mathcal{L}_{\mathrm{Pres}}$-definable function is piece-wise linear:

THEOREM 8.13 (Presburger cell decomposition [**Clu03**, Theorem 1]). *Let $X \subseteq \mathbb{Z}^m$ and let $f : X \to \mathbb{Z}$ be $\mathcal{L}_{\mathrm{Pres}}$-definable. Then there exists a finite partition*

$$X = \bigsqcup_{i=1}^{N} A_i,$$

such that the restriction $f|_{A_i} : A_i \to \mathbb{Z}$ is linear for each $1 \leq i \leq N$.

We further have,

THEOREM 8.14 (Rectilinearization, see [**Clu03**, Theorem 2]). *Let $X \subseteq \mathbb{Z}^m$ be $\mathcal{L}_{\mathrm{Pres}}$-definable set. Then there exists a finite partition*

$$X = \bigsqcup_{i=1}^{N} A_i,$$

and linear bijections $f_i : A_i \to \mathbb{N}^{l_i}$ for each i.

Another useful property of the model-theoretic formalism, is that certain properties of motivic functions $g = (g_F)_{F \in \mathrm{Loc}}$, such as integrability and boundedness of g_F, depend only on the isomorphism class of the residue field k_F of F, if the residual characteristic is large enough.

THEOREM 8.15 (Transfer principle for bounds, [**CGH16**, Theorem 3.1]). *Let X be an $\mathcal{L}_{\mathrm{DP}}$-definable set, and let $H, G \in \mathcal{C}(X)$ be motivic functions. Then the following holds for $F \in \mathrm{Loc}_{\gg}$; if*

$$|H_F(x)| \leq |G_F(x)|,$$

for all $x \in X_F$, then for any $F' \in \mathrm{Loc}$ with the same residue field as F, one also has

$$|H_{F'}(x)| \leq |G_{F'}(x)|,$$

for all $x \in X_{F'}$.

8.4. Number theoretic characterization of ε-jet flatness

Let $q = p^r$ be a prime power. We denote by $\mathbb{Q}_q$ the unique unramified extension of $\mathbb{Q}_p$ of degree r, its ring of integers by $\mathbb{Z}_q$, and the maximal ideal of $\mathbb{Z}_q$ by $\mathfrak{m}_q$.

Given an $\mathcal{O}_F$-scheme X, we have a natural reduction map $r_k : X(\mathcal{O}_F) \to X(\mathcal{O}_F/\mathfrak{m}_F^k)$. If $\varphi : X \to Y$ is an $\mathcal{O}_F$-morphism between $\mathcal{O}_F$-varieties X and Y, by abuse of notation we denote the natural maps $X(\mathcal{O}_F/\mathfrak{m}_F^k) \to Y(\mathcal{O}_F/\mathfrak{m}_F^k)$ by φ as well, and so for any $\overline{y} \in Y(\mathcal{O}_F/\mathfrak{m}_F^k)$, $\varphi^{-1}(\overline{y})$ is a finite set in $X(\mathcal{O}_F/\mathfrak{m}_F^k)$.

DEFINITION 8.16. Let X be a variety defined over a local field F. A *Schwartz measure* μ on $X(F)$ is a compactly supported, smooth measure (i.e. μ is locally a Haar measure in the analytic topology on $X(F)$).

LEMMA 8.17 (see e.g. [**Wei82**, Theorem 2.25] and [**AA18**, Proposition 3.1.1, arXiv version]). *If X is a smooth variety over $\mathcal{O}_F$ of pure relative dimension d, then there is a unique Schwartz measure, denoted $\mu_{X(\mathcal{O}_F)}$, such that for every $k \in \mathbb{N}$ and $\bar{x} \in X(\mathcal{O}_F/\mathfrak{m}_F^k)$,*

$$\mu_{X(\mathcal{O}_F)}(r_k^{-1}(\bar{x})) = |\mathcal{O}_F/\mathfrak{m}_F|^{-kd}.$$

We now prove the following theorem:

THEOREM 8.18. *Let $\varphi : X \to Y$ be a dominant morphism between finite type $\mathbb{Z}$-schemes X and Y, where $X_{\mathbb{Q}}$ and $Y_{\mathbb{Q}}$ are smooth and geometrically irreducible, and let $0 < \varepsilon \leq 1$. Then the following are equivalent:*

(1) *For any $0 < \varepsilon' < \varepsilon$ there exists a finite set S of primes, such that for every $q \in \mathcal{P}_S$, $k \in \mathbb{N}$ and $y \in Y(\mathbb{Z}_q/\mathfrak{m}_q^k)$,*

$$\left|\varphi^{-1}(y)\right| < q^{k(\dim X_{\mathbb{Q}} - \varepsilon' \dim Y_{\mathbb{Q}})}.$$

(2) *$\varphi_{\mathbb{Q}} : X_{\mathbb{Q}} \to Y_{\mathbb{Q}}$ is ε-jet flat.*

REMARK 8.19. If we restrict Condition (1) of Theorem 8.18 to $k = 1$ (i.e. when $\mathbb{Z}_q/\mathfrak{m}_q^k \simeq \mathbb{F}_q$), then $\left|\varphi^{-1}(y)\right| < q^{(\dim X_{\mathbb{Q}} - \varepsilon' \dim Y_{\mathbb{Q}})}$ is equivalent to $\varphi_{\mathbb{Q}} : X_{\mathbb{Q}} \to Y_{\mathbb{Q}}$ being ε-flat. This follows directly by observing that the fibers of φ are of bounded complexity, and applying the Lang-Weil bounds. Since the complexity of the fibers of $J_k(\varphi_{\mathbb{Q}})$ is arbitrarily large for $k \gg 0$, Theorem 8.18 deserves a special treatment.

For the proof of Theorem 8.18 we need some preparation. Let $\varphi : X \to Y$ be as in the theorem, assume that Y is affine and let $\mu_{X(\mathcal{O}_F)}$ and $\mu_{Y(\mathcal{O}_F)}$ be the measures as in Lemma 8.17 for $F \in \mathrm{Loc}_{\gg}$. Since φ is dominant it follows that $\varphi_*(\mu_{X(\mathcal{O}_F)})$ is absolutely continuous with respect to $\mu_{Y(\mathcal{O}_F)}$, and thus has an L^1-density (see e.g. [**AA16**, Corollary 3.6]). Set $\tau_F := \varphi_*(\mu_{X(\mathcal{O}_F)})$, and write $\tau_F = f_F(y)\cdot\mu_{Y(\mathcal{O}_F)}$. Moreover, for any $k\in\mathbb{N}$ define a function

$$g_F(y,k) = \frac{1}{\mu_{Y(\mathcal{O}_F)}(B(y,k))}\int_{B(y,k)} f_F(\widetilde{y})d\mu_{Y(\mathcal{O}_F)},$$

where $B(y,k) = r_k^{-1}(r_k(y))$. By Theorem 8.10 and Remark 8.11 (see also [**GH19**, Theorem 4.2]), it follows that both $(f_F : Y(F)\to\mathbb{C})_{F\in\mathrm{Loc}}$ and $(g_F : Y(F)\times\mathbb{N}\to\mathbb{C})_{F\in\mathrm{Loc}}$ are $\mathcal{L}_{\mathrm{DP}}$-motivic functions. Note that for any $F\in\mathrm{Loc}_{\gg}$, any $y\in Y(\mathcal{O}_F)$ and any $k\in\mathbb{N}$

$$g_F(y,k) = \frac{\left|\varphi_*(\mu_{X(\mathcal{O}_F)})(B(y,k)\right|}{\mu_{Y(\mathcal{O}_F)}(B(y,k))} = \frac{\left|\varphi^{-1}(r_k(y))\right|}{q^{k(\dim X_{\mathbb{Q}}-\dim Y_{\mathbb{Q}})}}. \tag{8.2}$$

PROOF OF THEOREM 8.18. (1) $\Rightarrow$ (2). By Theorem 8.15 and by possibly enlarging S, we have for each $k\in\mathbb{N}$ and each $y\in Y(\mathbb{F}_q[t]/t^k)$ that

$$\left|\varphi^{-1}(y)\right| < q^{k(\dim X_{\mathbb{Q}}-\varepsilon'\dim Y_{\mathbb{Q}})}.$$

Identify y with $\tilde{y}\in J_{k-1}(Y)(\mathbb{F}_q)$ via the identification $Y(\mathbb{F}_q[t]/t^k)\simeq J_{k-1}(Y)(\mathbb{F}_q)$. Note that

$$\left|\varphi^{-1}(y)\right| = \left|(J_{k-1}(X))_{\tilde{y},J_{k-1}(\varphi)}(\mathbb{F}_q)\right|.$$

By the Lang-Weil bounds, this implies that $\varphi_{\mathbb{Q}}$ is ε-jet flat.

(2) $\Rightarrow$ (1). Assume that $\varphi_{\mathbb{Q}}$ is ε-jet flat. Fix $\varepsilon'<\varepsilon$. By (8.2) it is enough to show that $g_F(y,k)q_F^{-k(1-\varepsilon')\dim Y_{\mathbb{Q}}} < 1$ for $F\in\mathrm{Loc}_{\gg}$. Note that for each individual k, this follows by applying the Lang-Weil bounds to $J_{k-1}(\varphi)$, so it is enough to show this for k large enough (depending on the choice of ϵ'). By [**CGH18**, Theorem 2.1.3], there exist a motivic function G on $\mathbb{N}$ and $d\in\mathbb{Z}$, such that for any $F\in\mathrm{Loc}_{\gg}$

$$\sup_{y\in Y(F)} g_F(y,k) \le G_F(k) \le q_F^d \sup_{y\in Y(F)} g_F(y,k).$$

It is therefore enough to show that for k large enough, and $F\in\mathrm{Loc}_{\gg}$,

$$H_F(k) := G_F(k)q_F^{-k(1-\varepsilon')\dim Y_{\mathbb{Q}}} < 1. \tag{$\star$}$$

LEMMA 8.20. *There exist a definable partition* $\mathbb{N} = \bigsqcup_{j=1}^{N} A_j$ *of* $\mathbb{N}$, *natural numbers* $L, \{a_i\}_{i=1}^L \in\mathbb{N}$ *and* $\{b_i\}_{i=1}^L\in\mathbb{Q}$, *such that on each definable part*

$$G_F(k) = \sum_{i=1}^{L} c_i(F)\cdot k^{a_i} q_F^{b_i k},$$

for any $F\in\mathrm{Loc}_{\gg}$ *and some constants* $\{c_i(F)\}_{i=1}^L$ *depending on* F.

PROOF. Since $G_F(k)$ is a motivic function, by Definition 8.9, it can be written as

$$G_F(k) = \sum_{i=1}^{N_1} |Y_{i,F,k}|\, q_F^{\alpha_{i,F}(k)} \left(\prod_{j=1}^{N_2}\beta_{ij,F}(k)\right)\left(\prod_{j=1}^{N_3}\frac{1}{1-q_F^{a_{ij}}}\right), \tag{$\star\star$}$$

where $\alpha_i:\mathbb{N}\to\mathbb{Z}$ and $\beta_{ij}:\mathbb{N}\to\mathbb{Z}$ are definable functions and $Y_i\subseteq\mathbb{N}\times\mathrm{RF}^{r_i}$ are definable sets. By the disjointness of the RF and VG sorts (see e.g. [**CL08**, Theorem 2.1.1]), we may partition $\mathbb{N}$ into finitely many definable subsets, such that on each part $Y_{i,F,k}=Y_{i,F}$ does not depend on k, and α_i,β_{ij} are $\mathcal{L}_{\mathrm{Pres}}$-definable functions.

By further refining the partition, we may assume that α_i, β_{ij} are linear (Theorem 8.13). By rearranging $(\star\star)$, the lemma follows. □

It is enough to prove $(\star)$ for each definable subset A_j, so by the last lemma we may assume that

$$H_F(k) = \sum_{i=1}^{L} c_i(F) \cdot k^{a_i} q_F^{k(b_i-(1-\varepsilon')\dim Y)}.$$

By our assumption that $k \gg 1$, it is enough to consider the case where A_j is infinite. Without loss of generality we may assume that b_1 is maximal. Set $T := \{i \in [L] : b_i = b_1\}$ and $P_F(k) := \sum_{i\in T} c_i(F) \cdot k^{a_i}$. Then we can write

$$H_F(k) = P_F(k) q_F^{k(b_1-(1-\varepsilon')\dim Y_{\mathbb{Q}})} + E_F(k),$$

where

$$E_F(k) = \sum_{i\in T^c} c_i(F) \cdot k^{a_i} q_F^{k(b_i-(1-\varepsilon')\dim Y_{\mathbb{Q}})}.$$

We may assume that for any $N \in \mathbb{N}$ there exists $F \in \mathrm{Loc}$ with $\mathrm{char}(k_F) > N$ such that $P_F(k)$ is not identically zero as a polynomial on k. Denote by Loc_P the collection of such $F \in \mathrm{Loc}$, and set $\mathcal{S}_P = \{q_F : F \in \mathrm{Loc}_P\}$, and $\mathrm{Loc}_{q,P} = \{F \in \mathrm{Loc}_P : q_F = q\}$.

LEMMA 8.21.

(1) *There exists $k_0 \in \mathbb{N}$ such that for any $k > k_0$, there are infinitely many $q \in \mathcal{S}_P$, such that $\exists F \in \mathrm{Loc}_{q,P}$ with $P_F(k) \neq 0$.*
(2) *There exists $e \in \mathbb{N}$ such that if $P_F(k) \neq 0$ for some $k \in \mathbb{N}$, then $q_F^{-e} < |P_F(k)|$ and moreover, $c_i(F) < q_F^e$ for each $i \in L$ and each $q_F > 2$.*
(3) *For any $\delta > 0$ and any $a \in \mathbb{N}$ there exists $k_1 \in \mathbb{N}$ such that $k^a < q^{\delta k}$ for every $k > k_1$ and every $q \geq 2$.*

PROOF. (1) Assume the contrary, and set $d := \deg(P)$. Then for any $k_0 \in \mathbb{N}$, there exists $k' > k_0$ such that $P_F(k') = 0$ for any $F \in \mathrm{Loc}_{q,P}$ with $q \in \mathcal{S}_P$ large enough. By induction, this implies that for q_F large enough, $P_F(k)$ has more than d roots yielding a contradiction.

Item (2) follows by analyzing the possible terms appearing in a motivic function (e.g. $(\star\star)$). Item (3) is obvious. □

COROLLARY 8.22. *There exists $k_2 \in \mathbb{N}$ such that for each $k > k_2$,*

$$\lim_{q_F\to\infty} P_F(k) q_F^{k(b_1-(1-\varepsilon')\dim Y_{\mathbb{Q}})} = 0,$$

and $P_F(k) \neq 0$ for infinitely many $q \in \mathcal{S}_P$.

PROOF. Let $\delta = \frac{1}{2}\min_{i\in T^c} |b_1 - b_i|$. By the last lemma, for any k large enough and $q_F > 2$, we have

$$|E_F(k)| < \sum_{i\in T^c} q_F^{e+k(\delta+b_i-(1-\varepsilon')\dim Y_{\mathbb{Q}})}.$$

Hence, there exists $k_2 \in \mathbb{N}$ such that for every $k > k_2$ and any $F \in \mathrm{Loc}_{q,P}$ with $P_F(k) \neq 0$, we have

$$|E_F(k)| < \frac{1}{2}\left|P_F(k) q_F^{k(b_1-(1-\varepsilon')\dim Y_{\mathbb{Q}})}\right|. \tag{8.3}$$

By Lemma 8.21(1), we may choose k_2 large enough such that for any $k > k_2$, we have $P_F(k) \neq 0$ for infinitely many $q \in \mathcal{S}_P$ and $F \in \mathrm{Loc}_{q,P}$. Since $\lim\limits_{q_F \to \infty} H_F(k) = 0$ for any k large enough (by the Lang-Weil bounds), and by (8.3) we are done. □

Corollary 8.22 implies the following:

COROLLARY 8.23. *We have* $b_1 \leq (1-\varepsilon)\mathrm{dim} Y_{\mathbb{Q}}$.

For any $\varepsilon' < \varepsilon$ we can now write $H_F(k) = \sum\limits_{i=1}^{L} c_i(F) \cdot k^{a_i} q_F^{kd_i(\varepsilon')}$, with all $d_i(\varepsilon') < 0$. Set $d(\varepsilon') = \max\limits_{1 \leq i \leq L} \{d_i(\varepsilon')\}$, then by Lemma 8.21 for every $\delta > 0$ we have k_0 such that for every $k > k_0$

$$H_F(k) < q_F^{(d(\varepsilon')+\delta)k+e}.$$

In particular, there exists k_1 such that $H_F(k) < 1$ for every $k > k_1$ and every $F \in \mathrm{Loc}_{\gg}$. Repeating this process for all cells $A_1, \ldots, A_N$, finishes the proof. □

CHAPTER 9

Applications to the p-adic probabilistic Waring type problem

Our goal in this section is to combine our results on word maps (Sections 3-7), with the number-theoretic interpretation of the (FRS) property and of ε-jet flatness (Section 8), to obtain results on random walks induced from word maps. We start by discussing the more general case of random walks induced from algebraic morphisms, and then descend to the case of word maps.

9.1. Algebraic random walks: General case

Let X be a finite type $\mathbb{Z}$-scheme, with $X_{\mathbb{Q}}$ smooth and geometrically irreducible, let $\underline{G}$ be a finite type affine group scheme, with $\underline{G}_{\mathbb{Q}}$ connected and let $\varphi : X \to \underline{G}$ be a dominant morphism.

Let $\mu_{X(\mathbb{Z}_q/\mathfrak{m}_q^k)}$ denote the uniform probability measure on $X(\mathbb{Z}_q/\mathfrak{m}_q^k)$. For every prime power q and $k \in \mathbb{N}$, the map $\varphi : X(\mathbb{Z}_q/\mathfrak{m}_q^k) \to \underline{G}(\mathbb{Z}_q/\mathfrak{m}_q^k)$ gives rise to a probability measure $\tau_{\varphi,q,k} := \varphi_*(\mu_{X(\mathbb{Z}_q/\mathfrak{m}_q^k)})$ by considering the pushforward of $\mu_{X(\mathbb{Z}_q/\mathfrak{m}_q^k)}$ under φ. Explicitly, the probability given to $g \in \underline{G}(\mathbb{Z}_q/\mathfrak{m}_q^k)$ is

$$\tau_{\varphi,q,k}(g) = \frac{\left|\varphi^{-1}(g)\right|}{\left|X(\mathbb{Z}_q/\mathfrak{m}_q^k)\right|}.$$

The collection $\{\tau_{\varphi,q,k}\}_{q,k}$ induces a family of random walks on $\{\underline{G}(\mathbb{Z}_q/\mathfrak{m}_q^k)\}_{q,k}$. Note that $\tau_{\varphi,q,k}^{*2} = \tau_{\varphi^{*2},q,k}$ (see e.g. [**GH19**, Remark 1.2]), so the t-th step of the random walk induced by $\tau_{\varphi,q,k}$ is given by

$$\tau_{\varphi,q,k}^{*t}(g) = \frac{\left|(\varphi^{*t})^{-1}(g)\right|}{\left|X(\mathbb{Z}_q/\mathfrak{m}_q^k)\right|^t}.$$

Let $\pi_{q,k}$ be the uniform probability on $\underline{G}(\mathbb{Z}_q/\mathfrak{m}_q^k)$. Recall that if $\operatorname{supp}(\tau_{\varphi,q,k})$ is generating and contains the identity element, then $\tau_{\varphi,q,k}^{*t}$ converges to $\pi_{q,k}$ as $t \to \infty$, and the mixing time (see Definition 1.12) measures this convergence rate. Also recall we define L^a norms as follows:

DEFINITION 9.1. For a real function f on a finite group G, set

$$\|f\|_a := (|G|^{a-1} \sum_{g\in G} |f(g)|^a)^{1/a} \text{ and } \|f\|_\infty = |G| \cdot \max_{g\in G} |f(g)|.$$

The following is a uniform analogue of mixing time.

DEFINITION 9.2. Let $\varphi : X \to \underline{G}$ be a dominant morphism as above. We say φ

(1) is *almost $\{\mathbb{F}_q\}$-uniform in L^a* if there exists a finite set S of primes such that

$$\lim_{\mathcal{P}_S \ni q \to \infty} \left\| \tau_{\varphi,q,1} - \pi_{q,1} \right\|_a = 0.$$

(2) has an *almost $\{\mathbb{F}_q\}$-uniform L^a-mixing time* $t_{a,\varphi,\{\mathbb{F}_q\}}$ if it is minimal such that $\varphi^{*t_{a,\varphi,\{\mathbb{F}_q\}}}$ is almost $\{\mathbb{F}_q\}$-uniform in L^a.
(3) is *almost $\{\mathbb{Z}_q\}$-uniform in L^a* if there exists a finite set S of primes such that

$$\lim_{\mathcal{P}_S \ni q \to \infty} \left(\sup_{k \in \mathbb{N}} \left\| \tau_{\varphi,q,k} - \pi_{q,k} \right\|_a \right) = 0.$$

(4) has an *almost $\{\mathbb{Z}_q\}$-uniform L^a-mixing time* $t_{a,\varphi,\{\mathbb{Z}_q\}}$ if it is minimal such that $\varphi^{*t_{a,\varphi,\{\mathbb{Z}_q\}}}$ is almost $\{\mathbb{Z}_q\}$-uniform in L^a.

9.1.1. Algebraic interpretation of almost uniform mixing time. We now use results from Chapter 8 to give an algebraic interpretation of various properties of the family of random walks induced by a morphism $\varphi : X \to \underline{G}$ as above. We first characterize properties related to L^∞ bounds and then deal with properties related to L^1 bounds.

THEOREM 9.3 (Characterization of L^∞-bounds). *Let X be a finite type $\mathbb{Z}$-scheme, with $X_{\mathbb{Q}}$ smooth and geometrically irreducible, let $\underline{G}$ be a finite type affine group scheme, with $\underline{G}_{\mathbb{Q}}$ connected and let $\varphi : X \to \underline{G}$ be a dominant morphism.*

(1) *φ is almost $\{\mathbb{F}_q\}_q$-uniform in L^∞ if and only if $\varphi_{\mathbb{Q}} : X_{\mathbb{Q}} \to \underline{G}_{\mathbb{Q}}$ is (FGI).*
(2) *φ is almost $\{\mathbb{Z}_q\}_q$-uniform in L^∞ if and only if $\varphi_{\mathbb{Q}} : X_{\mathbb{Q}} \to \underline{G}_{\mathbb{Q}}$ is (FRS) and (FGI).*
(3) *$\varphi_{\mathbb{Q}}$ is ε-jet flat if and only if for every $\varepsilon' < \varepsilon$ there exists a set $S(\varepsilon')$ of primes such that the family $\{\tau_{\varphi,q,k}\}_{q,k}$ satisfies $\tau_{\varphi,q,k}(g) \leq \left|\underline{G}(\mathbb{Z}_q/\mathfrak{m}_q^k)\right|^{-\varepsilon'}$ for every $q \in \mathcal{P}_S$ and every $k \in \mathbb{Z}_{>0}$.*

PROOF.

(1) Assume that $\varphi_{\mathbb{Q}} : X_{\mathbb{Q}} \to \underline{G}_{\mathbb{Q}}$ is (FGI). Theorem 8.4 implies that φ is almost $\{\mathbb{F}_q\}_q$-uniform in L^∞. For the other direction, Theorem 8.4 implies that $\varphi_{\mathbb{Q}}$ is flat. Assume that $\varphi_{\mathbb{Q}} : X_{\mathbb{Q}} \to \underline{G}_{\mathbb{Q}}$ is not (FGI). By [**Gro67**, Corollary 9.7.9], we may find a prime p, such that for each $r \in \mathbb{N}$ and $q = p^r$ the map $\varphi_{\mathbb{F}_q} : X_{\mathbb{F}_q} \to \underline{G}_{\mathbb{F}_q}$ is flat but not (FGI). Take $g \in \underline{G}(\mathbb{F}_q)$ with $C_{\varphi,q,g} > 1$. Then also $C_{\varphi,q^{r'},g} > 1$ for any $r' \in \mathbb{N}$. By Theorem 8.4 we deduce that φ is not almost $\{\mathbb{F}_q\}_q$-uniform in L^∞.
(2) Assume that φ is almost $\{\mathbb{Z}_q\}_q$-uniform in L^∞. Item (1) implies that $\varphi_{\mathbb{Q}}$ is (FGI). By enlarging S we may assume that X is smooth over $\mathbb{Z}_q$ for every $q \in \mathcal{P}_S$, so that $\mu_{X(\mathbb{Z}_q)}$ and $\mu_{\underline{G}(\mathbb{Z}_q)}$ exist by Lemma 8.17. Since φ is dominant, $\varphi_*(\mu_{X(\mathbb{Z}_q)})$ is absolutely continuous with respect to $\mu_{\underline{G}(\mathbb{Z}_q)}$ (see e.g. [**AA16**, Corollary 3.6]). Fix a prime $p \notin S$, and let $q = p^r \in \mathcal{P}_S$. For

every $g \in \underline{G}(\mathbb{Z}_q)$ set $B(g,k) = r_k^{-1}(r_k(g))$, then we have,

$$(\dagger) \qquad \frac{\mu_{X(\mathbb{Z}_q)}(\varphi^{-1}(B(g,k)))}{\mu_{\underline{G}(\mathbb{Z}_q)}(B(g,k))} = \frac{|\varphi^{-1}(r_k(g))|}{q^{k(\dim X_{\mathbb{Q}} - \dim \underline{G}_{\mathbb{Q}})}} = \tau_{\varphi,q,k}(r_k(g)) \cdot \frac{|X(\mathbb{Z}_q/\mathfrak{m}_q^k)|}{q^{k(\dim X_{\mathbb{Q}} - \dim \underline{G}_{\mathbb{Q}})}} \leq \sup_{k\in\mathbb{N}} \|\tau_{\varphi,q,k}\|_\infty \frac{|X(\mathbb{Z}_q/\mathfrak{m}_q^k)|}{q^{k\dim X_{\mathbb{Q}}}} \frac{q^{k\dim \underline{G}_{\mathbb{Q}}}}{|\underline{G}(\mathbb{Z}_q/\mathfrak{m}_q^k)|}.$$

Since φ is almost $\{\mathbb{Z}_q\}_q$-uniform in L^∞, by using the the Lang-Weil bounds, and by possibly enlarging S, it follows that ($\dagger$) is uniformly bounded (in k) by a constant $C > 0$. By the analytic criterion for the (FRS) property (Theorem 8.5 and [**Gla19**, Theorem 1.6]) combined with [**GH19**, Lemma 3.15], the map $\varphi_{\mathbb{Q}}$ is (FRS).

For the converse, by [**CGH**, Theorem 4.7] one can find $C > 0$ such that for every $q \in \mathcal{P}_S$, every $k \in \mathbb{N}$ and every $g \in \underline{G}(\mathbb{Z}_q/\mathfrak{m}_q^k)$, one has:

$$\left| \frac{\tau_{\varphi,q,k}(g) \cdot |X(\mathbb{Z}_q/\mathfrak{m}_q^k)|}{q^{k(\dim X_{\mathbb{Q}} - \dim \underline{G}_{\mathbb{Q}})}} - 1 \right| = \left| \frac{|\varphi^{-1}(g)|}{q^{k(\dim X_{\mathbb{Q}} - \dim \underline{G}_{\mathbb{Q}})}} - 1 \right| < C \cdot q^{-1/2}.$$

By applying the Lang-Weil estimates, and since $X_{\mathbb{Q}}$ and $\underline{G}_{\mathbb{Q}}$ are geometrically irreducible, one can find $C' > 0$ such that, by possibly enlarging S, one has

$$\|\tau_{\varphi,q,k} - \pi_{q,k}\|_\infty < C' \cdot q^{-1/2}, \tag{9.1}$$

for every $q \in \mathcal{P}_S$ and every $k \in \mathbb{N}$. Hence, φ is almost $\{\mathbb{Z}_q\}_q$-uniform in L^∞.

(3) Follows from Theorem 8.18 by using the Lang-Weil bounds. □

The following theorem is a strengthening of [**LST19**, Proposition 2.1].

THEOREM 9.4 (Characterization of L^1-uniformity). *Let $\varphi : X \to \underline{G}$ be as in Theorem 9.3. Then the following are equivalent:*

(1) *φ is almost $\{\mathbb{F}_q\}_q$-uniform in L^1.*
(2) *φ is almost $\{\mathbb{Z}_q\}_q$-uniform in L^1.*
(3) *$\varphi_{\mathbb{Q}} : X_{\mathbb{Q}} \to \underline{G}_{\mathbb{Q}}$ has a geometrically irreducible generic fiber.*

PROOF. We would like to prove (1) ⇒ (3) ⇒ (2) ⇒ (1). The implication (2) ⇒ (1) is obvious. (1) ⇒ (3) is essentially given in [**LST19**]: if φ is almost $\{\mathbb{F}_q\}_q$-uniform in L^1, then for any prime $p \notin S$ we have $\lim_{n\to\infty} \|\tau_{\varphi,p^n} - \pi_{p^n}\|_1 = 0$. By [**LST19**, Proposition 2.1], $\varphi_{\mathbb{F}_p}$ has geometrically irreducible generic fiber for each $p \notin S$, which implies that $\varphi_{\mathbb{Q}}$ has a geometrically irreducible generic fiber.

It is left to show (3) ⇒ (2). Let $U \subseteq \underline{G}$ be an open set such that $\varphi_{\mathbb{Q}}$ is smooth and (FGI) over $U_{\mathbb{Q}} \subseteq \underline{G}_{\mathbb{Q}}$. Set

$$V_q := \{g \in \underline{G}(\mathbb{Z}_q) : r_1(g) \in U(\mathbb{F}_q)\}.$$

By a generalization of Hensel's lemma, we may find a finite set S such that for every $q \in \mathcal{P}_S$, and every $g \in V_q$ we have

$$|\varphi^{-1}(r_k(g))| = q^{(k-1)(\dim X_{\mathbb{Q}} - \dim \underline{G}_{\mathbb{Q}})} \cdot |\varphi^{-1}(r_1(g))|.$$

By the Lang-Weil bounds, we may find $C_1 > 0$ such that for any $g \in V_q$ we have

$$\left| \tau_{\varphi,q,k}(r_k(g)) - \frac{1}{\left|\underline{G}(\mathbb{Z}_q/\mathfrak{m}_q^k)\right|} \right| < \frac{C_1 q^{-1/2}}{\left|\underline{G}(\mathbb{Z}_q/\mathfrak{m}_q^k)\right|}.$$

Write $Z = U^c$ and notice that $\dim \varphi^{-1}(Z)_{\mathbb{Q}} < \dim X_{\mathbb{Q}}$ since $X_{\mathbb{Q}}$ is geometrically irreducible. We may therefore find $C_2 > 0$, such that

$$\begin{aligned} \|\tau_{\varphi,q,k} - \pi_{q,k}\|_1 &= \sum_{g \in r_k(V_q)} |\tau_{\varphi,q,k}(g) - \pi_{q,k}(g)| + \sum_{g \in r_k(V_q^c)} |\tau_{\varphi,q,k}(g) - \pi_{q,k}(g)| \\ &< C_1 q^{-1/2} + \sum_{g \in r_k(V_q^c)} |\tau_{\varphi,q,k}(g)| + \sum_{g \in r_k(V_q^c)} |\pi_{q,k}(g)| \\ &< C_1 q^{-1/2} + \frac{\left|\varphi^{-1}(Z(\mathbb{F}_q))\right|}{|X(\mathbb{F}_q)|} + \frac{|Z(\mathbb{F}_q)|}{|\underline{G}(\mathbb{F}_q)|} \\ &< C_1 q^{-1/2} + C_2 q^{-1}, \end{aligned}$$

where the last inequality follows by applying the Lang-Weil bounds. □

We would now like to discuss the random walk on $\underline{G}(\hat{\mathbb{Z}}) = \varprojlim \underline{G}(\mathbb{Z}/n\mathbb{Z})$. We say that a map $\varphi : X \to \underline{G}$ is *almost* $\{\mathbb{Z}/n\mathbb{Z}\}_{n\in\mathbb{N}}$*-uniform in* L^a if there exists a set S of primes such that $\|\tau_{\varphi,n} - \pi_n\|_a < \frac{1}{2}$ for every $n \in \mathbb{N}$ which is a product of primes outside of S. By the Chinese remainder theorem, it is tempting to say that if $\varphi : X \to \underline{G}$ is almost $\{\mathbb{Z}_q\}$-uniform in L^∞, then it is also *almost* $\{\mathbb{Z}/n\mathbb{Z}\}_{n\in\mathbb{N}}$*-uniform in* L^∞. The following example shows this hope is too naive.

Remark 9.5 (Warning! Rational singularities+complete intersection does not imply good point count over $\mathbb{Z}/n\mathbb{Z}$). In fact, even smooth schemes might have bad point count over $\mathbb{Z}/n\mathbb{Z}$; let E be an elliptic curve over $\mathbb{Q}$. The Sato-Tate conjecture (see [**CHT08, Tay08, HSBT10**]) says that under some mild assumptions on E, we have $|E(\mathbb{F}_p)| = p + 1 - 2\sqrt{p}\cos(\theta_p)$, where for any $0 \leq \alpha \leq \beta \leq \pi$,

$$\lim_{M\to\infty} \left(\frac{\#\{\text{primes } p \leq M : \alpha \leq \theta_p \leq \beta\}}{\#\{\text{primes } p \leq M\}} \right) = \frac{2}{\pi} \int_\alpha^\beta \sin^2(\theta) d\theta. \tag{9.2}$$

In particular, by (9.2) and the prime number theorem, there exist $\varepsilon, \delta > 0$ such that for arbitrary large $M \in \mathbb{N}$, we may find a subset S of primes in the interval $[\frac{1}{2}M, M]$, such that $|S| > \varepsilon \left| \frac{M}{\log M} \right|$ and such that $\frac{|E(\mathbb{F}_p)|}{p} \geq 1 + \delta p^{-1/2}$, for every $p \in S$. Now, taking $n = \prod_{p\in S} p$, by the Chinese remainder theorem we get the following:

$$\frac{|E(\mathbb{Z}/n\mathbb{Z})|}{n} = \prod_{p\in S} \frac{|E(\mathbb{F}_p)|}{p} \geq \prod_{p\in S} \left(1 + \delta p^{-1/2}\right) \geq \left(1 + \frac{\delta}{\sqrt{M}}\right)^{\varepsilon \left|\frac{M}{\log M}\right|} \overset{M\to\infty}{\longrightarrow} \infty.$$

It turns out that this pathology can be fixed by applying a few more convolutions.

Proposition 9.6. *Let* $\varphi : X \to \underline{G}$ *be as in Theorem 9.3. Assume that* $\varphi_{\mathbb{Q}}$ *is (FRS) and (FGI). Then* φ^{*3} *is almost* $\{\mathbb{Z}/n\mathbb{Z}\}_{n\in\mathbb{N}}$*-uniform in* L^∞.

PROOF. By (9.1), there exists $C > 0$, such that for every $k \in \mathbb{N}$ and almost every prime p

$$\|\tau_{\varphi,p,k} - \pi_{p,k}\|_{\infty} < Cp^{-\frac{1}{2}}.$$

By standard properties of mixing (see e.g. [**LP17**, Lemma 4.18]), we have

$$\left\|\tau^{*t}_{\varphi,p,k} - \pi_{p,k}\right\|_{\infty} < C^t p^{-\frac{t}{2}}.$$

Now let $n, N \in \mathbb{N}$ be such that

$$n = \prod_{i=1}^{N} p_i^{k_i}$$

with p_i distinct primes, larger than M for some $M \in \mathbb{N}$ sufficiently large. Then for each i and each $g \in \underline{G}(\mathbb{Z}/p_i^{k_i}\mathbb{Z})$,

$$1 - C^t p_i^{-\frac{t}{2}} < \left|\underline{G}(\mathbb{Z}/p_i^{k_i}\mathbb{Z})\right| \tau^{*t}_{\varphi,p_i,k_i}(g) < 1 + C^t p_i^{-\frac{t}{2}}.$$

Therefore for every $g \in \underline{G}(\mathbb{Z}/n\mathbb{Z})$, every $t \geq 3$, and M sufficiently large we get,

$$\begin{aligned}
|\underline{G}(\mathbb{Z}/n\mathbb{Z})| \, \tau^{*t}_{\varphi,n}(g) &< \exp\left(\sum_{M<p\in \text{primes}} \log\left(1 + C^t p^{-\frac{t}{2}}\right)\right) \\
&< \exp\left(\sum_{M<p\in \text{primes}} C^t p^{-\frac{t}{2}}\right) \\
&< 1 + \frac{1}{2},
\end{aligned}$$

and similarly

$$1 - \frac{1}{2} < |\underline{G}(\mathbb{Z}/n\mathbb{Z})| \, \tau^{*t}_{\varphi,n}(g),$$

which is precisely what we want. □

REMARK 9.7. The (FRS) property can also be used to obtain information on algebraic random walks on real algebraic groups. Indeed, an Archimedean analogue of [**AA16**, Theorem 3.4] was proven in [**Rei**]: the pushforward $\varphi_*(\mu)$ of any smooth compactly supported measure μ on $X(\mathbb{R})$ under an (FRS) morphism $\varphi : X \to Y$ between smooth $\mathbb{R}$-varieties is a measure with continuous density.

9.2. Algebraic random walks induced from word maps

Let $\mathcal{G}$ denote the collection of all affine groups schemes $\underline{G}$, with $\underline{G}_{\mathbb{Q}}$ semisimple and simply connected. We study uniform properties of the family of random walks induced by a word $w \in F_r$ on $\{\underline{G}(\mathbb{Z}_q)\}_{q,\underline{G}\in\mathcal{G}}$ by considering the family of random walks induced on the quotients. Let $\pi_{\underline{G},q,k}$ be the uniform probability measure on $\underline{G}(\mathbb{Z}_q/\mathfrak{m}_q^k)$. The word map φ_w induces a collection of probability measures $\{\tau_{w,\underline{G},q,k}\}_{\underline{G},q,k}$ on $\{\underline{G}(\mathbb{Z}_q/\mathfrak{m}_q^k)\}_{\underline{G},q,k}$ by $\tau_{w,\underline{G},q,k} := (\varphi_w)_*(\pi^r_{\underline{G},q,k})$. As we saw before, the probability to reach an element $g \in \underline{G}(\mathbb{Z}_q/\mathfrak{m}_q^k)$ at the t-th step of the random walk is given by

$$\tau^{*t}_{w,\underline{G},q,k}(g) = \frac{\left|(\varphi_w^{*t})^{-1}(g)\right|}{\left|\underline{G}(\mathbb{Z}_q/\mathfrak{m}_q^k)\right|^{rt}}.$$

DEFINITION 9.8. Let $w \in F_r$ be a word.

(1) w is *almost* $\{\mathbb{Z}_q\}_q$*-uniform in* L^a if $\varphi_w : \underline{G}^r \to \underline{G}$ is almost $\{\mathbb{Z}_q\}_q$-uniform for every $\underline{G} \in \mathcal{G}$.
(2) w has an *almost* $\{\mathbb{Z}_q\}_q$*-uniform* L^a*-mixing time* $t_{a,w,\{\mathbb{Z}_q\}_q}$ if it is minimal such that $w^{*t_{a,w,\{\mathbb{Z}_q\}_q}}$ is almost-$\{\mathbb{Z}_q\}_q$-uniform in L^a.

These notions are similarly defined for the collection $\{\mathbb{F}_q\}_q$.

The following corollary is an immediate consequence of Theorems 9.3 and 9.4.

COROLLARY 9.9. *Let* $w \in F_r$.
(1) *w is almost $\{\mathbb{F}_q\}$-uniform in L^∞ if and only if $(\varphi_w)_\mathbb{Q} : \underline{G}^r_\mathbb{Q} \to \underline{G}_\mathbb{Q}$ is (FGI) for every $\underline{G} \in \mathcal{G}$.*
(2) *w is almost $\{\mathbb{Z}_q\}$-uniform in L^∞ if and only if $(\varphi_w)_\mathbb{Q} : \underline{G}^r_\mathbb{Q} \to \underline{G}_\mathbb{Q}$ is (FRS) and (FGI) for every $\underline{G} \in \mathcal{G}$.*
(3) *The following are equivalent:*
  (a) *w is almost $\{\mathbb{F}_q\}$-uniform in L^1.*
  (b) *w is almost $\{\mathbb{Z}_q\}$-uniform in L^1.*
  (c) *$(\varphi_w)_\mathbb{Q} : \underline{G}^r_\mathbb{Q} \to \underline{G}_\mathbb{Q}$ has a geometrically irreducible generic fiber, for every $\underline{G} \in \mathcal{G}$.*

9.2.1. Uniform properties of random walks induced by word measures. Combining the discussion on random walks in the last two subsections, together with the results on word maps obtained in this paper, we can summarize our applications to p-adic probabilistic Waring type problem.

The following is a direct consequence of Theorem 7.9 and Item (3) of Theorem 9.3.

COROLLARY 9.10. *Let $w \in F_r$ be a word. Then for any $n \in \mathbb{N}$ there exists a finite set $S(n)$ of primes such that for every $q \in \mathcal{P}_{S(n)}$ and every $g \in \mathrm{SL}_n(\mathbb{Z}_q/\mathfrak{m}_q^k)$,*

$$\tau_{w,\mathrm{SL}_n,q,k}(g) \leq \left|\mathrm{SL}_n(\mathbb{Z}_q/\mathfrak{m}_q^k)\right|^{-\frac{1}{2\cdot 10^{17}\ell(w)^9}}.$$

Theorem 7.8 allows us to provide L^∞-bounds for random walks on congruence subgroups. For any $\underline{G} \in \mathcal{G}$ and any prime p, write $\underline{G}^1(\mathbb{Z}_p)$ for the kernel of the quotient map $\underline{G}(\mathbb{Z}_p) \to \underline{G}(\mathbb{F}_p)$.

COROLLARY 9.11. *Let $w \in F_r$ be a word. Then there exists $C > 0$, such that for every $\underline{G} \in \mathcal{G}$, every prime $p > p_0(\underline{G})$, and every $t \geq C\ell(w)^4$, the pushforward measure $(\varphi_w^{*t})_*(\mu^{rt}_{\underline{G}^1(\mathbb{Z}_p)})$ on $\underline{G}^1(\mathbb{Z}_p)$ has continuous density with respect to the normalized Haar measure.*

PROOF. By Theorem 7.8, there exists $U \subseteq \underline{G}^{rt}$ such that $\varphi_w^{*t} : U_\mathbb{Q} \to \underline{G}_\mathbb{Q}$ is (FRS) for $t = C\ell(w)^4$. Note that for $p \gg 1$, $U(\mathbb{F}_p)$ contains $(e, \ldots, e) \in \underline{G}(\mathbb{F}_p)^{rt}$ and U is smooth over $\mathbb{Z}_p$. By a generalization of Hensel's lemma, we have $U(\mathbb{Z}_p) \supseteq \underline{G}^1(\mathbb{Z}_p)^{rt}$. By the analytic criterion for the (FRS) property (Theorem 8.5), $(\varphi_w^{*t})_*(\mu^{rt}_{\underline{G}^1(\mathbb{Z}_p)})$ has continuous density. □

COROLLARY 9.12 (L^1-mixing time, generalization of [**LST19**, Theorems 1 and 2]). *Let $w_1 \in F_{r_1}$ and $w_2 \in F_{r_2}$ be words. Then $w = w_1 * w_2$ is almost $\{\mathbb{Z}_q\}_q$-uniform in L^1.*

PROOF. By Theorem 3.19, $(\varphi_w)_\mathbb{Q}$ has geometrically irreducible generic fiber for every $\underline{G} \in \mathcal{G}$. By Corollary 9.9 we are done. □

COROLLARY 9.13 (Mixing time for the commutator map). *Let* $w_0 = xyx^{-1}y^{-1} \in F_r$ *be the commutator word. Then*

(1) w_0 *is almost* $\{\mathbb{Z}_q\}_q$*-uniform in* L^1.
(2) w_0^{*4} *is almost* $\{\mathbb{Z}_q\}_q$*-uniform in* L^∞.

PROOF. By Theorems 6.1, 6.2 and by Corollary 3.20, the map $\varphi_{w_0}^{*4} : \underline{G}^8 \to \underline{G}$ is (FRS) and (FGI) for every $\underline{G} \in \mathcal{G}$. By **[GS09]**, w_0 is $\{\mathbb{F}_q\}$*-uniform in* L^1. By Corollary 9.9 we are done. $\square$

Bibliography

[AA16] Avraham Aizenbud and Nir Avni, *Representation growth and rational singularities of the moduli space of local systems*, Invent. Math. **204** (2016), no. 1, 245–316, DOI 10.1007/s00222-015-0614-8. MR3480557

[AA18] Avraham Aizenbud and Nir Avni, *Counting points of schemes over finite rings and counting representations of arithmetic lattices*, Duke Math. J. **167** (2018), no. 14, 2721–2743, DOI 10.1215/00127094-2018-0021. MR3859363

[AGKS13] Nir Avni, Tsachik Gelander, Martin Kassabov, and Aner Shalev, *Word values in p-adic and adelic groups*, Bull. Lond. Math. Soc. **45** (2013), no. 6, 1323–1330, DOI 10.1112/blms/bdt063. MR3138499

[AL50] A. S. Amitsur and J. Levitzki, *Minimal identities for algebras*, Proc. Amer. Math. Soc. **1** (1950), 449–463, DOI 10.2307/2032312. MR36751

[AM19] Nir Avni and Chen Meiri, *Words have bounded width in* $\mathrm{SL}(n, \mathbb{Z})$, Compos. Math. **155** (2019), no. 7, 1245–1258, DOI 10.1112/s0010437x19007334. MR3963488

[BGKP12] Tatiana Bandman, Nikolai Gordeev, Boris Kunyavskiĭ, and Eugene Plotkin, *Equations in simple Lie algebras*, J. Algebra **355** (2012), 67–79, DOI 10.1016/j.jalgebra.2012.01.012. MR2889532

[BLST18] Roman Bezrukavnikov, Martin W. Liebeck, Aner Shalev, and Pham Huu Tiep, *Character bounds for finite groups of Lie type*, Acta Math. **221** (2018), no. 1, 1–57, DOI 10.4310/ACTA.2018.v221.n1.a1. MR3877017

[Bor83] A. Borel, *On free subgroups of semisimple groups*, Enseign. Math. (2) **29** (1983), no. 1-2, 151–164. MR702738

[Bor17] Alexander Bors, *Fibers of automorphic word maps and an application to composition factors*, J. Group Theory **20** (2017), no. 6, 1103–1133, DOI 10.1515/jgth-2017-0024. MR3719319

[Bud21] Nero Budur, *Rational singularities, quiver moment maps, and representations of surface groups*, Int. Math. Res. Not. IMRN **15** (2021), 11782–11817, DOI 10.1093/imrn/rnz236. MR4294133

[CB01] William Crawley-Boevey, *Geometry of the moment map for representations of quivers*, Compositio Math. **126** (2001), no. 3, 257–293, DOI 10.1023/A:1017558904030. MR1834739

[CGH] Raf Cluckers, Itay Glazer, and Yotam I. Hendel, *A number theoretic characterization of E-smooth and (FRS) morphisms: estimates on the number of* $\mathbb{Z}/p^k\mathbb{Z}$*-points*, Algebra Number Theory **17** (2023), no. 12, 2229–2260, DOI 10.2140/ant.2023.17.2229. MR4650394

[CGH14] Raf Cluckers, Julia Gordon, and Immanuel Halupczok, *Integrability of oscillatory functions on local fields: transfer principles*, Duke Math. J. **163** (2014), no. 8, 1549–1600, DOI 10.1215/00127094-2713482. MR3210968

[CGH16] Raf Cluckers, Julia Gordon, and Immanuel Halupczok, *Transfer principles for bounds of motivic exponential functions*, Families of automorphic forms and the trace formula, Simons Symp., Springer, [Cham], 2016, pp. 111–127. MR3675165

[CGH18] Raf Cluckers, Julia Gordon, and Immanuel Halupczok, *Uniform analysis on local fields and applications to orbital integrals*, Trans. Amer. Math. Soc. Ser. B **5** (2018), 125–166, DOI 10.1090/btran/25. MR3859937

[Chi06] E. S. Chibrikov, *A right normed basis for free Lie algebras and Lyndon-Shirshov words*, J. Algebra **302** (2006), no. 2, 593–612, DOI 10.1016/j.jalgebra.2006.03.036. MR2293773

[CHT08] Laurent Clozel, Michael Harris, and Richard Taylor, *Automorphy for some l-adic lifts of automorphic mod l Galois representations*, Publ. Math. Inst. Hautes Études Sci. **108** (2008), 1–181, DOI 10.1007/s10240-008-0016-1. With Appendix A, summarizing unpublished work of Russ Mann, and Appendix B by Marie-France Vignéras. MR2470687

[Chu76] V. N. Čubarikov, *Multiple rational trigonometric sums and multiple integrals* (Russian), Mat. Zametki **20** (1976), no. 1, 61–68. MR429785

[CL08] Raf Cluckers and François Loeser, *Constructible motivic functions and motivic integration*, Invent. Math. **173** (2008), no. 1, 23–121, DOI 10.1007/s00222-008-0114-1. MR2403394

[CL10] Raf Cluckers and François Loeser, *Constructible exponential functions, motivic Fourier transform and transfer principle*, Ann. of Math. (2) **171** (2010), no. 2, 1011–1065, DOI 10.4007/annals.2010.171.1011. MR2630060

[CLNS18] Antoine Chambert-Loir, Johannes Nicaise, and Julien Sebag, *Motivic integration*, Progress in Mathematics, vol. 325, Birkhäuser/Springer, New York, 2018, DOI 10.1007/978-1-4939-7887-8. MR3838446

[Clu03] Raf Cluckers, *Presburger sets and p-minimal fields*, J. Symbolic Logic **68** (2003), no. 1, 153–162, DOI 10.2178/jsl/1045861509. MR1959315

[Clu04] Raf Cluckers, *Multivariate Igusa theory: decay rates of exponential sums*, Int. Math. Res. Not. **76** (2004), 4093–4108, DOI 10.1155/S1073792804141913. MR2109987

[CX18] Serge Cantat and Junyi Xie, *Algebraic actions of discrete groups: the p-adic method*, Acta Math. **220** (2018), no. 2, 239–295, DOI 10.4310/ACTA.2018.v220.n2.a2. MR3849285

[Den87] J. Denef, *On the degree of Igusa's local zeta function*, Amer. J. Math. **109** (1987), no. 6, 991–1008, DOI 10.2307/2374583. MR919001

[Elk78] Renée Elkik, *Singularités rationnelles et déformations* (French), Invent. Math. **47** (1978), no. 2, 139–147, DOI 10.1007/BF01578068. MR501926

[EM04] Lawrence Ein and Mircea Mustaţă, *Inversion of adjunction for local complete intersection varieties*, Amer. J. Math. **126** (2004), no. 6, 1355–1365. MR2102399

[EM09] Lawrence Ein and Mircea Mustaţă, *Jet schemes and singularities*, Algebraic geometry—Seattle 2005. Part 2, Proc. Sympos. Pure Math., vol. 80, Part 2, Amer. Math. Soc., Providence, RI, 2009, pp. 505–546, DOI 10.1090/pspum/080.2/2483946. MR2483946

[ET15] Abdelrhman Elkasapy and Andreas Thom, *On the length of the shortest non-trivial element in the derived and the lower central series*, J. Group Theory **18** (2015), no. 5, 793–804, DOI 10.1515/jgth-2015-0007. MR3393415

[GH19] Itay Glazer and Yotam I. Hendel, *On singularity properties of convolutions of algebraic morphisms*, Selecta Math. (N.S.) **25** (2019), no. 1, Paper No. 15, 41, DOI 10.1007/s00029-019-0457-z. MR3911739

[GH21] Itay Glazer and Yotam I. Hendel, *On singularity properties of convolutions of algebraic morphisms—the general case*, J. Lond. Math. Soc. (2) **103** (2021), no. 4, 1453–1479, DOI 10.1112/jlms.12414. With an appendix by Glazer, Hendel and Gady Kozma. MR4273476

[Gla19] Itay Glazer, *On rational singularities and counting points of schemes over finite rings*, Algebra Number Theory **13** (2019), no. 2, 485–500, DOI 10.2140/ant.2019.13.485. MR3927053

[GLT] Robert M. Guralnick, Michael Larsen, and Pham Huu Tiep, *Character levels and character bounds*, Forum Math. Pi **8** (2020), e2, 81, DOI 10.1017/fmp.2019.9. MR4061963

[GLT20] Robert M. Guralnick, Michael Larsen, and Pham Huu Tiep, *Character levels and character bounds for finite classical groups*, Invent. Math. **235** (2024), no. 1, 151–210, DOI 10.1007/s00222-023-01221-5. MR4688703

[Gro67] A. Grothendieck, *Éléments de géométrie algébrique. IV. Étude locale des schémas et des morphismes de schémas IV* (French), Inst. Hautes Études Sci. Publ. Math. **32** (1967), 361. MR238860

[GS09] Shelly Garion and Aner Shalev, *Commutator maps, measure preservation, and T-systems*, Trans. Amer. Math. Soc. **361** (2009), no. 9, 4631–4651, DOI 10.1090/S0002-9947-09-04575-9. MR2506422

[Hil09] David Hilbert, *Beweis für die Darstellbarkeit der ganzen Zahlen durch eine feste Anzahl n^{ter} Potenzen (Waringsches Problem)* (German), Math. Ann. **67** (1909), no. 3, 281–300, DOI 10.1007/BF01450405. MR1511530

[Hir64] Heisuke Hironaka, *Resolution of singularities of an algebraic variety over a field of characteristic zero. I, II*, Ann. of Math. (2) **79** (1964), 109–203; ibid. (2) **79** (1964), 205–326, DOI 10.2307/1970547. MR0199184

[HLS15] Chun Yin Hui, Michael Larsen, and Aner Shalev, *The Waring problem for Lie groups and Chevalley groups*, Israel J. Math. **210** (2015), no. 1, 81–100, DOI 10.1007/s11856-015-1246-9. MR3430269

[HSBT10] Michael Harris, Nick Shepherd-Barron, and Richard Taylor, *A family of Calabi-Yau varieties and potential automorphy*, Ann. of Math. (2) **171** (2010), no. 2, 779–813, DOI 10.4007/annals.2010.171.779. MR2630056

[JZ08] Andrei Jaikin-Zapirain, *On the verbal width of finitely generated pro-p groups*, Rev. Mat. Iberoam. **24** (2008), no. 2, 617–630, DOI 10.4171/RMI/549. MR2459206

[Kap] Guy Kapon, *Singularity properties of graph varieties*, `arXiv:1905.05847` (2019).

[Kas07] Martin Kassabov, *Universal lattices and unbounded rank expanders*, Invent. Math. **170** (2007), no. 2, 297–326, DOI 10.1007/s00222-007-0064-z. MR2342638

[Lar04] Michael Larsen, *Word maps have large image*, Israel J. Math. **139** (2004), 149–156, DOI 10.1007/BF02787545. MR2041227

[Li93] Jun Li, *The space of surface group representations*, Manuscripta Math. **78** (1993), no. 3, 223–243, DOI 10.1007/BF02599310. MR1206154

[LOST10] Martin W. Liebeck, E. A. O'Brien, Aner Shalev, and Pham Huu Tiep, *The Ore conjecture*, J. Eur. Math. Soc. (JEMS) **12** (2010), no. 4, 939–1008, DOI 10.4171/JEMS/220. MR2654085

[LP17] David A. Levin and Yuval Peres, *Markov chains and mixing times*, American Mathematical Society, Providence, RI, 2017. Second edition of [MR2466937]; With contributions by Elizabeth L. Wilmer; With a chapter on "Coupling from the past" by James G. Propp and David B. Wilson, DOI 10.1090/mbk/107. MR3726904

[LS01] Martin W. Liebeck and Aner Shalev, *Diameters of finite simple groups: sharp bounds and applications*, Ann. of Math. (2) **154** (2001), no. 2, 383–406, DOI 10.2307/3062101. MR1865975

[LS08] Michael Larsen and Aner Shalev, *Characters of symmetric groups: sharp bounds and applications*, Invent. Math. **174** (2008), no. 3, 645–687, DOI 10.1007/s00222-008-0145-7. MR2453603

[LS09] Michael Larsen and Aner Shalev, *Word maps and Waring type problems*, J. Amer. Math. Soc. **22** (2009), no. 2, 437–466, DOI 10.1090/S0894-0347-08-00615-2. MR2476780

[LS12] Michael Larsen and Aner Shalev, *Fibers of word maps and some applications*, J. Algebra **354** (2012), 36–48, DOI 10.1016/j.jalgebra.2011.10.040. MR2879221

[LST11] Michael Larsen, Aner Shalev, and Pham Huu Tiep, *The Waring problem for finite simple groups*, Ann. of Math. (2) **174** (2011), no. 3, 1885–1950, DOI 10.4007/annals.2011.174.3.10. MR2846493

[LST19] Michael Larsen, Aner Shalev, and Pham Huu Tiep, *Probabilistic Waring problems for finite simple groups*, Ann. of Math. (2) **190** (2019), no. 2, 561–608, DOI 10.4007/annals.2019.190.2.3. MR3997129

[LW54] Serge Lang and André Weil, *Number of points of varieties in finite fields*, Amer. J. Math. **76** (1954), 819–827, DOI 10.2307/2372655. MR65218

[MKS76] Wilhelm Magnus, Abraham Karrass, and Donald Solitar, *Combinatorial group theory*, Second revised edition, Dover Publications, Inc., New York, 1976. Presentations of groups in terms of generators and relations. MR0422434

[MP10] Justin Malestein and Andrew Putman, *On the self-intersections of curves deep in the lower central series of a surface group*, Geom. Dedicata **149** (2010), 73–84, DOI 10.1007/s10711-010-9465-z. MR2737679

[MSS18] Laurentiu Maxim, Morihiko Saito, and Jörg Schürmann, *Thom-Sebastiani theorems for filtered $\mathcal{D}$-modules and for multiplier ideals*, Int. Math. Res. Not. IMRN **1** (2020), 91–111, DOI 10.1093/imrn/rny032. MR4050564

[Mus01] Mircea Mustaţă, *Jet schemes of locally complete intersection canonical singularities*, Invent. Math. **145** (2001), no. 3, 397–424, DOI 10.1007/s002220100152. With an appendix by David Eisenbud and Edward Frenkel. MR1856396

[Mus02] Mircea Mustaţă, *Singularities of pairs via jet schemes*, J. Amer. Math. Soc. **15** (2002), no. 3, 599–615, DOI 10.1090/S0894-0347-02-00391-0. MR1896234

[Mus12] Mircea Mustaţă, *IMPANGA lecture notes on log canonical thresholds*, Contributions to algebraic geometry, EMS Ser. Congr. Rep., Eur. Math. Soc., Zürich, 2012, pp. 407–442, DOI 10.4171/114-1/16. Notes by Tomasz Szemberg. MR2976952

[MZ96] C. Martinez and E. Zelmanov, *Products of powers in finite simple groups*. part B, Israel J. Math. **96** (1996), no. part B, 469–479, DOI 10.1007/BF02937318. MR1433702

[Raz73] Ju. P. Razmyslov, *The existence of a finite basis for the identities of the matrix algebra of order two over a field of characteristic zero* (Russian), Algebra i Logika **12** (1973), 83–113, 121. MR0340348

[Rei] Andrew Reiser, *Pushforwards of measures on real varieties under maps with rational singularities*, arXiv:1807.00079 (2018).

[Ric79] R. W. Richardson, *Commuting varieties of semisimple Lie algebras and algebraic groups*, Compositio Math. **38** (1979), no. 3, 311–327. MR535074

[Sel67] G. B. Seligman, *Modular Lie algebras*, Ergebnisse der Mathematik und ihrer Grenzgebiete, Band 40, Springer-Verlag New York, Inc., New York, 1967. MR0245627

[Sha09] Aner Shalev, *Word maps, conjugacy classes, and a noncommutative Waring-type theorem*, Ann. of Math. (2) **170** (2009), no. 3, 1383–1416, DOI 10.4007/annals.2009.170.1383. MR2600876

[SS11] Gili Schul and Aner Shalev, *Words and mixing times in finite simple groups*, Groups Geom. Dyn. **5** (2011), no. 2, 509–527, DOI 10.4171/GGD/137. MR2782183

[Sta18] The Stacks Project Authors. *Stacks Project*. https://stacks.math.columbia.edu, 2018.

[Stu91] Garrett Stuck, *Low-dimensional actions of semisimple groups*, Israel J. Math. **76** (1991), no. 1-2, 27–71, DOI 10.1007/BF02782843. MR1177331

[SW97] Jan Saxl and John S. Wilson, *A note on powers in simple groups*, Math. Proc. Cambridge Philos. Soc. **122** (1997), no. 1, 91–94, DOI 10.1017/S030500419600165X. MR1443588

[Tay08] Richard Taylor, *Automorphy for some l-adic lifts of automorphic mod l Galois representations. II*, Publ. Math. Inst. Hautes Études Sci. **108** (2008), 183–239, DOI 10.1007/s10240-008-0015-2. MR2470688

[Wei82] André Weil, *Adeles and algebraic groups*, Progress in Mathematics, vol. 23, Birkhäuser, Boston, Mass., 1982. With appendices by M. Demazure and Takashi Ono. MR670072

Editorial Information

To be published in the *Memoirs*, a paper must be correct, new, nontrivial, and significant. Further, it must be well written and of interest to a substantial number of mathematicians. Piecemeal results, such as an inconclusive step toward an unproved major theorem or a minor variation on a known result, are in general not acceptable for publication.

Papers appearing in *Memoirs* are generally at least 80 and not more than 200 published pages in length. Papers less than 80 or more than 200 published pages require the approval of the Managing Editor of the Transactions/Memoirs Editorial Board. Published pages are the same size as those generated in the style files provided for $\mathcal{A}_\mathcal{M}\mathcal{S}$-LaTeX.

Information on the backlog for this journal can be found on the AMS website starting from `http://www.ams.org/memo`.

A Consent to Publish is required before we can begin processing your paper. After a paper is accepted for publication, the Providence office will send a Consent to Publish and Copyright Agreement to all authors of the paper. By submitting a paper to the *Memoirs*, authors certify that the results have not been submitted to nor are they under consideration for publication by another journal, conference proceedings, or similar publication.

Information for Authors

Memoirs is an author-prepared publication. Once formatted for print and on-line publication, articles will be published as is with the addition of AMS-prepared frontmatter and backmatter. Articles are not copyedited; however, confirmation copy will be sent to the authors.

Initial submission. The AMS uses Centralized Manuscript Processing for initial submissions. Authors should submit a PDF file using the Initial Manuscript Submission form found at `www.ams.org/submission/memo`, or send one copy of the manuscript to the following address: Centralized Manuscript Processing, MEMOIRS OF THE AMS, 201 Charles Street, Providence, RI 02904-2294 USA. If a paper copy is being forwarded to the AMS, indicate that it is for *Memoirs* and include the name of the corresponding author, contact information such as email address or mailing address, and the name of an appropriate Editor to review the paper (see the list of Editors below).

The paper must contain a *descriptive title* and an *abstract* that summarizes the article in language suitable for workers in the general field (algebra, analysis, etc.). The *descriptive title* should be short, but informative; useless or vague phrases such as "some remarks about" or "concerning" should be avoided. The *abstract* should be at least one complete sentence, and at most 300 words. Included with the footnotes to the paper should be the 2020 *Mathematics Subject Classification* representing the primary and secondary subjects of the article. The classifications are accessible from `www.ams.org/msc/`. The Mathematics Subject Classification footnote may be followed by a list of *key words and phrases* describing the subject matter of the article and taken from it. Journal abbreviations used in bibliographies are listed in the latest *Mathematical Reviews* annual index. The series abbreviations are also accessible from `www.ams.org/msnhtml/serials.pdf`. To help in preparing and verifying references, the AMS offers MR Lookup, a Reference Tool for Linking, at `www.ams.org/mrlookup/`.

Electronically prepared manuscripts. The AMS encourages electronically prepared manuscripts, with a strong preference for $\mathcal{A}_\mathcal{M}\mathcal{S}$-LaTeX. To this end, the Society has prepared $\mathcal{A}_\mathcal{M}\mathcal{S}$-LaTeX author packages for each AMS publication. Author packages include instructions for preparing electronic manuscripts, samples, and a style file that generates the particular design specifications of that publication series.

Authors may retrieve an author package for *Memoirs of the AMS* from `www.ams.org/journals/memo/memoauthorpac.html`. The *AMS Author Handbook* is available in PDF format from the author package link. The author package can also be obtained free of charge by sending email to `tech-support@ams.org` or from the Publication Division,

American Mathematical Society, 201 Charles St., Providence, RI 02904-2294, USA. When requesting an author package, please specify the publication in which your paper will appear. Please be sure to include your complete mailing address.

After acceptance. The source files for the final version of the electronic manuscript should be sent to the Providence office immediately after the paper has been accepted for publication. The author should also submit a PDF of the final version of the paper to the editor, who will forward a copy to the Providence office.

Accepted electronically prepared files can be submitted via the web at `www.ams.org/submit-book-journal/`, sent via FTP, or sent on CD to the Electronic Prepress Department, American Mathematical Society, 201 Charles Street, Providence, RI 02904-2294 USA. TEX source files and graphic files can be transferred over the Internet by FTP to the Internet node `ftp.ams.org` (130.44.1.100). When sending a manuscript electronically via CD, please be sure to include a message indicating that the paper is for the *Memoirs*.

Electronic graphics. Comprehensive instructions on preparing graphics are available at `www.ams.org/authors/journals.html`. A few of the major requirements are given here.

Submit files for graphics as EPS (Encapsulated PostScript) files. This includes graphics originated via a graphics application as well as scanned photographs or other computer-generated images. If this is not possible, TIFF files are acceptable as long as they can be opened in Adobe Photoshop or Illustrator.

Authors using graphics packages for the creation of electronic art should also avoid the use of any lines thinner than 0.5 points in width. Many graphics packages allow the user to specify a "hairline" for a very thin line. Hairlines often look acceptable when proofed on a typical laser printer. However, when produced on a high-resolution laser imagesetter, hairlines become nearly invisible and will be lost entirely in the final printing process.

Screens should be set to values between 15% and 85%. Screens which fall outside of this range are too light or too dark to print correctly. Variations of screens within a graphic should be no less than 10%.

Any graphics created in color will be rendered in grayscale for the printed version unless color printing is authorized by the Managing Editor and the Publisher. In general, color graphics will appear in color in the online version.

Inquiries. Any inquiries concerning a paper that has been accepted for publication should be sent to `memo-query@ams.org` or directly to the Electronic Prepress Department, American Mathematical Society, 201 Charles St., Providence, RI 02904-2294 USA.

SELECTED PUBLISHED TITLES IN THIS SERIES

1486 **M. Gekhtman, M. Shapiro, and A. Vainshtein,** A Plethora of Cluster Structures on GL_n, 2024

1485 **David Ayala, Aaron Mazel-Gee, and Nick Rozenblyum,** Stratified Noncommutative Geometry, 2024

1484 **Brandon Doherty, Krzysztof Kapulkin, Zachery Lindsey, and Christian Sattler,** Cubical Models of $(\infty, 1)$-Categories, 2024

1483 **Masao Oi,** Simple Supercuspidal L-Packets of Quasi-Split Classical Groups, 2024

1482 **David Burns and Daniel Macias Castillo,** On Refined Conjectures of Birch and Swinnerton-Dyer Type for Hasse–Weil–Artin L-Series, 2024

1481 **Sonja Cox, Martin Hutzenthaler, and Arnulf Jentzen,** Local Lipschitz Continuity in the Initial Value and Strong Completeness for Nonlinear Stochastic Differential Equations, 2024

1480 **Nicolas Burq and Laurent Thomann,** Almost Sure Scattering for the One Dimensional Nonlinear Schrödinger Equation, 2024

1479 **Eva Elduque, Christian Geske, Moisés Herradón Cueto, Laurenţiu G. Maxim, and Botong Wang,** Mixed Hodge Structures on Alexander Modules, 2024

1478 **Qi Chen, Dongyi Wei, and Zhifei Zhang,** Transition Threshold for the 3D Couette Flow in a Finite Channel, 2024

1477 **Vincent Guirardel, Gilbert Levitt, and Rizos Sklinos,** Towers and the First-Order Theories of Hyperbolic Groups, 2024

1476 **Gaëtan Borot, Vincent Bouchard, Nitin K. Chidambaram, Thomas Creutzig, and Dmitry Noshchenko,** Higher Airy Structures, $\mathcal{W}$ Algebras and Topological Recursion, 2024

1475 **Sam Chow and Niclas Technau,** Littlewood and Duffin–Schaeffer-Type Problems in Diophantine Approximation, 2024

1474 **Christian Blohmann and Alan Weinstein,** Hamiltonian Lie Algebroids, 2024

1473 **Dihua Jiang, Zhilin Luo, and Lei Zhang,** Harmonic Analysis and Gamma Functions on Symplectic Groups, 2024

1472 **Thomas Creutzig, Shashank Kanade, and Robert McRae,** Tensor Categories for Vertex Operator Superalgebra Extensions, 2024

1471 **Roberto Feola and Filippo Giuliani,** Quasi-Periodic Traveling Waves on an Infinitely Deep Perfect Fluid Under Gravity, 2024

1470 **Maxwell Stolarski,** Curvature Blow-up in Doubly-warped Product Metrics Evolving by Ricci Flow, 2024

1469 **Steve Boyer, Cameron McA. Gordon, and Xingru Zhang,** Dehn Fillings of Knot Manifolds Containing Essential Twice-Punctured Tori, 2024

1468 **Alice Hedenlund and John Rognes,** A Multiplicative Tate Spectral Sequence for Compact Lie Group Actions, 2024

1467 **Qi Lü and Xu Zhang,** Optimal Feedback for Stochastic Linear Quadratic Control and Backward Stochastic Riccati Equations in Infinite Dimensions, 2024

1466 **Martin W. Liebeck, Gary M. Seitz, and Donna M. Testerman,** Multiplicity-free Representations of Algebraic Groups, 2024

1465 **Tobias Rossmann and Christopher Voll,** Groups, Graphs, and Hypergraphs: Average Sizes of Kernels of Generic Matrices with Support Constraints, 2024

1464 **Jens Marklof and Andreas Strömbergsson,** Kinetic Theory for the Low-Density Lorentz Gas, 2024

1463 **You Qi and Joshua Sussan,** p-DG Cyclotomic nilHecke Algebras II, 2024

For a complete list of titles in this series, visit the
AMS Bookstore at **www.ams.org/bookstore/memoseries/**.